Online Communities and Open Innovation

The advent of Internet marked a significant change in how users and customers can be involved in the innovative process. History is rife with examples of how users innovate, but Internet and its associated communication technologies brought radically new means for individuals to interact rapidly and at little cost in communities that spur new innovations. These communities are initiated and governed by people that differ in their motivations for taking part and participate to varying degrees. Such communities are outside the immediate control of companies seeking to develop open innovation strategies aimed at harnessing their work. This book brings together distinguished scholars from different disciplines: economics, organization theory, innovation studies and marketing in order to provide an improved understanding of how technological, as well as symbolic, value is created and appropriated at the intersection between online communities and firms. Empirical examples are presented from different industries, including software, services and manufacturing. This book offers food for thought for academics and managers to an important phenomenon that challenges many conventional wisdoms for how business can be done.

This book was published as a special issue of *Industry and Innovation*.

Linus Dahlander is an Assistant Professor at the European School of Management and Technology, Berlin, Germany. His research investigates how new ideas and innovations are developed in networks and communities. The research seeks to understand how these communities and networks unfold over time when individuals are distributed and autonomous – self-selecting tasks and collaboration partners.

Lars Frederiksen works at the Innovation Management Group at Department of Marketing and Statistics, at Aarhus School of Business and Social Sciences, Denmark. His research explores the ideas, activities and roles of individuals and teams in innovative projects, communities and organizations. He studies the development of knowledge creation, integration and transfer through social networks. Lars employs both qualitative and quantitative methodologies.

Francesco Rullani is Assistant Professor in Entrepreneurship and Management of Innovation at LUISS Guido Carli, Department of Economics and Business, Rome, Italy. His research focuses on production of knowledge by self-organizing groups of actors. He has explored these themes in the field of free/open source software, employing mainly quantitative techniques applied to large databases.

Online Communities and Open Innovation

Governance and Symbolic Value Creation

Edited by
Linus Dahlander, Lars Frederiksen and Francesco Rullani

Routledge
Taylor & Francis Group
LONDON AND NEW YORK

First published 2011
by Routledge
4 Park Square, Milton Park, Abingdon, Oxon OX14 4RN

Simultaneously published in the USA and Canada
by Routledge
605 Third Avenue, New York, NY 10017

Routledge is an imprint of the Taylor & Francis Group, an informa business

This book is a reproduction of *Industry and Innovation* 15.2. The Publisher requests that those citing this book use the bibliographical details of the journal issue on which the book is based.

Typeset in Helvetica by Taylor & Francis Books

British Library Cataloguing in Publication Data
A catalogue record for this book is available from the British Library

ISBN13: 978-0-415-61748-2 (hbk)

Disclaimer
The publisher would like to make readers aware that the chapters in this book are referred to as articles as they had been in the special issue. The publisher accepts responsibility for any inconsistencies that may have arisen in the course of preparing this volume for print.

Contents

Online Communities and Open Innovation: Governance and Symbolic Value Creation

LINUS DAHLANDER, LARS FREDERIKSEN & FRANCESCO RULLANI

Introduction

How can firms make use of online communities as part of an innovation strategy aimed at leveraging resources and ideas outside the four walls of the enterprise? Online communities are today a widespread phenomenon that takes a variety of forms. Free and open source software is probably the most well-known case, where geographically dispersed individuals collectively develop new software and produce innovation. In 1991 Linus Torvalds founded the Linux kernel, the heart of an operating system with the ability to have a real impact on Microsoft's market share. Torvalds' initial ideas led to the building of a community that collectively developed the Linux kernel. From the original incorporation of some 10,000 lines of source code, by 2005 the community had developed more than 6,000,000 lines of code. But online communities are more than simply free and open source software. For instance, social networking sites such as Facebook and MySpace, which have memberships of millions, have grown rapidly, allowing individuals to share experiences and socialize with each other. From initially being exclusively for participation by Harvard students, Facebook, according to recent estimates, now has more than 60 million users worldwide. The popular press has been swift to document these successes, and it is tempting to conclude that online communities have great potential. Yet, their diversity, in terms of objectives, typology of organization, production and reasons behind individuals' use of them, is becoming obvious.

Using different theoretical lenses, a growing body of literature in social sciences attempts to unfold this phenomenon (see, e.g. recent special issues in *Research Policy*, 2003; *Management Science*, 2006; and *Organisation Studies*, 2007). Much of the research on online communities suggests that the nature of these communities with permeable boundaries and self-organization makes them a new powerful *locus* of collective creativity and innovation (Lee and Cole, 2003). This is because individuals distributed across the globe can work with other individuals with whom they have affinities and shared interests. There has been much less research on how firms use communities where they have limited control, which might even jeopardize their competitive advantage, as part of their business models.

In recent years, a growing number of papers has adopted the open innovation concept to highlight that innovation processes span organizational boundaries. As Chesbrough (2003: XXIV) puts it: "firms can and should use external ideas as well as internal ideas, and internal and external paths to market, as the firms look to advance their technology". This stream of research highlights that firms have to rely on ideas from the external environment to stay abreast of competition (Chesbrough, 2003). Research on open innovation focuses on situations where firm boundaries are permeable, resulting in an interactive and distributed innovation process in which ideas, resources and individuals flow in and out of organizations (Laursen and Salter, 2006). Some contributions to this line of thinking suggest that firms can adopt strategies to integrate ideas that originate from communities, into their internal processes (Dahlander and Wallin, 2006; Jeppesen and Frederiksen, 2006). Firms can act strategically, by selectively revealing internal resources (Henkel, 2006) or by establishing communities to enable individuals to start using their products (Jeppesen and Frederiksen, 2006).

Online communities, therefore, can constitute an important external source of innovation for those firms able to implement a constructive relationship with them (Dahlander and Magnusson, 2005). Individuals in these communities may not only be able to develop innovations that can be integrated into the firm, but also may come up with new perspectives on and ways of framing problems. The community may develop a shared and mutual understanding of what it is about, what in the new product design or features is valuable; it may create product/firm loyalty and establish among community participants a sense of belonging and meaning (Rindova and Petkova, 2007).

Despite these benefits, there is also a range of challenges for firms that adopt the open innovation approach (Chesbrough, 2006). This is particularly evident when managing online communities as individuals participating in these communities are beyond the firms' hierarchical realms. Individuals can decide where to work, who to work with and what to work on, making it difficult for firms to steer the direction of development (Dahlander and Wallin, 2006). Moreover, in online communities the social processes behind members' participation are intrinsically dissipative because in such self-organized processes, many individuals have to be mobilized to make the most productive ones emerge (David and Rullani, forthcoming). This greatly increases the resources firms have to pour into these communities, and increases the risk of such investments. A large number of involved parties with misaligned goals, different capabilities and diverse degrees of involvement, raise the issue of governance of online communities.

In order to advance our understanding of the open and distributed nature of the innovation process taking place through online communities, this Special Issue revolves around the two themes identified above as crucial: (1) the importance of conceptually including the symbolic value of the artefacts in the innovation process, as online communities can be fundamental tools by which firms can innovate in this sphere thickening the symbolic value of their product; and (2) the issue of governance and how it is associated with the way in which firms try to harness these communities. Both themes have been relatively unattended by earlier research. The papers in this issue were selected precisely on the basis of the questions and answers they might generate with respect to these overall themes.

We structure this editorial piece as follows. We begin by placing the collection of papers within the themes of governance and symbolic value creation. We then conclude by charting areas for future research.

Theme I: Governance in Online Communities

Many scholars are interested in how online communities are governed. Sometimes these communities emerge quickly and appear to be successful, without the reporting mechanisms and organization charts visible within firms (Fleming and Waguespack, 2007). Firms are often involved in online communities as the protagonists of a single community (Jeppesen and Frederiksen, 2006), hosts of a system composed by a myriad of project-centred communities (David and Rullani, 2008) or collaborators integrated in an existing community (Dahlander and Wallin, 2006).

Research on governance in online communities looks at how individuals in these communities make collective decisions about future directions, control and coordination (see, e.g. the discussion by Markus, 2007). Dealing with governance is not a simple matter, as it is clearly a composite and dynamic concept, based on the nexus between very heterogeneous actors (O'Mahony & Ferraro, 2007; Shah, 2006). The dimensions that it is necessary to take account of are several, and their links are not always clear.

The paper by Langlois and Garzarelli, "Of Hackers and Hairdressers: Modularity and the Organizational Economics of Open-Source Collaboration", is the first paper in this Special Issue and sets the stage for a discussion on governance in online communities, allowing us to tease out what are the important dimensions. In this mainly conceptual paper, the authors employ the empirical illustration of an open source online community to explore the generic question in organizational economics of how the division of intellectual labour is based on a trade-off between modularity (i.e. specialization) and the opportunity to integrate various individually developed components of knowledge. The paper claims that the trade-off allows the individuals populating the open source community to exchange efforts rather than products, under a regime in which the providers of code self-identify themselves as suppliers of products in a market, rather than employees in a firm. Through their discussion, Langlois and Garzarelli build a useful two-by-two matrix of product vs. efforts on one axis and self-identification of contributors vs. no self-identification on the other. In this matrix the firm, the market, outsourcing and voluntary production as it occurs in open source communities are situated and, hence, presented as different modes of innovation production.

This provokes a series of questions on how communities can be managed when the connection between incentives—that is, the voluntary basis upon which the community is built—and the particular dynamics of the organization of labour in an open community—exchanging effort and not product—is taken into account. Firms and communities have diverse and sometimes incommensurable goals (O'Mahony, 2003), and it is a challenge for firms to derive benefits from working with communities.

The West and O'Mahony paper, "The Role of Participation Architecture in Growing Sponsored Open Source Communities", offers an answer to the previous implicit question about governance structures and the contradictions of a series of open source communities classified according to the typologies of firms' participation in these communities. Based on a qualitative study the paper shows that firm-sponsored online communities or open source online communities initiated by a firm, differ from organically grown open source communities. To demonstrate the differences between these two archetypical forms of open source online communities West and O'Mahony develop the concept of "participation architecture". The concept is created by the joining together of three important design dimensions for the coordination of tasks and communication in an online community: management of intellectual property rights, development approach and model of community governance. The study makes it explicit that various participation architectures exist in the two kinds of open source communities. The authors find that corporate sponsorship in open source communities influences the design and evolution of them and that this affects: (1) the degree of transparency of community participants to follow the community's collective process of development; and (2) accessibility for participants, to contribute to code development. Despite oftentimes trying to imitate the organization and design of organic open source communities, firm-sponsored communities face the classic tension between control and growth. This is because firms that are sponsoring an open source community struggle to maintain an open structure supportive of growth in the community in parallel with managing and maintaining control over the direction of and the activities taking place in the open source online community. For example, a firm sponsoring a community may define and potentially limit the opportunity structure for others to enter the community, as well as deciding who has access to the code/core of the community. The final contribution of this paper to the debate invoked in this Special Issue demonstrates that it is rarely the technical architecture and set-up of online communities that single-handedly determines participation frequency and structure. To better understand the differences in the character and quality of participation in different types of online communities, and thus be better informed about how innovation through these communities is managed and incentivized, we need to note that the organizational structure hinges upon the community sponsor's decisions regarding the design of governance mechanisms.

The paper by den Besten and Dalle provides an opportunity to explore this issue from a different perspective, taking a very timely phenomenon as the empirical setting. Stretching the argument presented in the paper by West and O'Mahony, they offer an intuition on the relationship between organizational structure and governance of online communities: the coordination devices the community uses in its everyday work embodies, "expresses", its (implicit or explicit, consciously or unconsciously taken) decisions relative to governance. Evaluating the role of these mechanisms in the effectiveness of the community activity develops into an interesting exercise, as it suggests visions of new ways—and new mechanisms—by which firms can realize the effective governance of an online community.

The paper by den Besten and Dalle focuses on the exciting example of Wikipedia, the online encyclopaedia, developed at tremendous pace by a vibrant online community. From its inception in 2001 to the end of December 2007, the Wikipedia community had collectively written an astonishing 9 million plus articles, in over 200 languages. Interest has surged among organization and management theorists to understand how the community's efforts have been organized to develop at this pace. The paper studies Simple Wikipedia—an offspring of Wikipedia—which develops articles written in Basic English, aimed at children, people with learning disabilities and individuals who are not fully literate in English. Simple Wikipedia has a mechanism that allows members to label pages as "unsimple", to ensure readability. Using an impressive data-set of more than 25,000 pages written by almost 20,000 contributors, the paper investigates how this labelling influences the readability of articles. The paper adopts methods used in linguistics and information science, such as the Flesch formula, which measures the readability of articles, counting the number of syllables per word and the number of words per sentence in texts. The results suggest that labelling plays a key—yet insufficient—role in ensuring articles are readable. The implications, suggested by the authors, are that an artificial companion—configured as a managerial and editorial assistant—could be developed using the relatively straightforward metrics suggested in the paper. These companions could continue work left unattended by members of the community, and also free up time for them to do more creative tasks.

Because such mechanisms as the companion mentioned above, would have an impact on the content of the everyday work of contributors, it could also be used to shape the governance of the community. Drawing on the Simple Wikipedia example analysed by den Besten and Dalle, we can imagine a wider application of "bots" able to grease the coordination between contributors. Once the "bots" are set, only certain tasks will be left to human coordination and discussion—for instance, to confrontation of different interests—while other issues will be "hidden" and taken care of by the automatic procedures for specific tasks. This would have a great impact on the evolution of governance structure in online communities.

Theme II: Different Types of Value Creation in Online Communities

The literature has underlined the role of users as innovators. The usual perspective uncovers the role of lead users as those individuals whose everyday life is affected by the consumption of the firm's product, and who have the skills to modify and personalize the product. The innovation process, however, most of the time is described as "atomistic": the user tries to match her needs with the features of the product, and produces a modified version of it. Users in online communities are embedded in a social environment where they can get advice from others to solve particular problems. As suggested by the user innovation literature, communication plays a function as a propagation device, distributing innovative features or voicing needs in the user community and eventually carrying them to the firm. While being sympathetic to this argument, we would contend that communication among users is more than this. In other words, we need to look at the communication that was going on before, during and after Schumpeterian inventions and innovations proliferated, because that is the mechanism that joins together the two processes allowing invention and innovation to link up to produce the potential to become both creative *and* valuable.

In line with this perspective an emergent view on value creation in the context of innovation is that consumers or potential consumers are not only attracted to the functional aspects and features of products (Rindova and Petkova, 2007). Issues such as symbolic and aesthetic value as well as sense-making qualities seem equally important for creating value through successful product development (Ravasi and Rindova, 2007). The literature on user innovation and online communities, in our view, has frequently focused perhaps too narrowly on innovation as improvements to technological features and innovative task partitioning, and seldom recognized that innovation and invention in online communities also takes place through communication among participants and thus through the recombination of ideas, voicing of future product wants and convergence towards a common perception of what is valuable. Being part of the community and making sense of one's practices with the product by interacting with fellow users may be an important driver of innovative behaviour.

In a growing number of online communities where transparency in the actual product development process is relatively low and access to the product source code is restricted, the value created for the consumer lies as much in being part of the community, infusing the product with an extended meaning beyond the utilitarian one. As such, firms that in different ways are involved with online communities may benefit from the communication among product users and their sense-making of their activities with the product.

The contribution by Finotto and Di Maria in this Special Issue is an example for this reasoning. As the authors state: "Understanding and managing innovation in the present competitive scenario requires a stronger integration between innovation studies—traditionally focused on R&D and technological innovation in product development—and the analyses offered by the marketing and customer behaviour framework." They refer to the concept of brand communities to capture the process of value creation through communities. "A brand community is a specialized, non-geographically bound community, based on a structured set of social relationships among admirers of a brand" (Muniz and O'Guinn, 2001: 412). They initially review the literature in both fields and argue for a mixed approach. They then focus on brand communities and discuss how taking account of innovation in collective sense-making processes could become a fundamental element of firms' strategies. The core of their argument is that in mature sectors, where technological progress has reached a *plateau*, the *locus* of value creation resides mainly in the immaterial features of the product. They report fashion as the most important example of this process, where product innovation along dimensions such as the identity and lifestyle carried by the brand are the main determinants of consumers' choices. To give an empirical illustration of this process they report two case studies from low-tech firms exploiting consumer communities to deepen and broaden the sets of meanings coupled to their products. The two firms analysed are particularly interesting as they are among the most important representatives of the "Made in Italy" industries. Predominantly present in low-tech industrial districts, firms representing "Made in Italy" have always benefited from geographic proximity to the customers. Proximity was at the basis of the processes of information gathering, of adaptation to new trends and tastes. Nowadays, the changes induced by the widening of worldwide markets have exposed firms to a different, mostly unknown and heterogeneous, demand. As the authors argue, small Italian companies can overcome the sudden gap between them and these new markets by means of online communities. The two case studies they develop offer an interesting first step to thinking about possible strategies firms could apply to reach this goal.

The paper by Finotto and Di Maria focuses on the symbolic value of the product for the consumer. This, however, is just one side of the coin. In communities where consumers also play the role of innovators, the meanings carried by the product are not confined to the consumption sphere. Production becomes a matter not only of maximizing job opportunities, the fitting of the product to one's own needs or simply earning money. The drivers behind innovation include also intrinsic motivation, where the meaning of the action and the symbols the product embodies create the context that nurtures enjoyment and a sense of fulfilment.

The paper developed by Kaiser and Müller-Seitz, "Leveraging Lead User Knowledge in Software Development—The Case of Weblog Technology", helps to illustrate the point made above. It demonstrates that the same incentive schemes based on the coexistence of extrinsic and intrinsic motivation, which previous studies found in open source communities, can be found in firm-centred communities operating in the blogsphere.

The authors analyse the motivations driving individuals' participation in a peculiar online social environment: the Microsoft Longhorn Blogsphere (MLB). The authors explain "the blogsphere" as a network of messages posted online, where all refer to a specific topic, environment or interest and are connected by means of a series of interlinks (blogrolls, permalinks and trackbacks). MLB is a corporate blogsphere centred on discussion of bloggers' experiences and developments of the Microsoft product "Microsoft-Longhorn". Despite the peculiarities of the environment, the tools, and the typology of participants and contributions, the main message of the paper is that the bloggers' incentive structure is not so very different from that observed in open source communities. The paper makes this point through an accurate analysis of the contents of the messages posted on MLB, classifying them and matching the categories obtained to the literature on open source developers' incentives. They show that bloggers are motivated both intrinsically and extrinsically. In particular, intrinsic motivations are found to belong to three main "realms": (1) feelings of freedom to choose (topic, task and time allocation, etc.); (2) feelings of having an impact on the reality; (3) socialization. The authors highlight that the minimum common denominator of these three sets of motivations is the concept of "flow state", that is, an experience that totally captures an individual's attention and stimulates intrinsic feelings of enjoyment. Extrinsic motivations are summarized by the categories of signalling one's abilities, participation perceived as a means to get support for one's activities and writing on the blog as a self-reflective learning activity. These motivations are found to closely map those emerging from the empirical studies of open source software. This result leaves the door open for generalizing our discussion, as it shows that online communities share similar incentive structures irrespective of the specific means of interaction through which they develop or their "corporate-centred" nature.

Conclusion

This editorial has elucidated the richness of the online community research field when considered in the light of firm-based open innovation strategy. The selection of papers for this Special Issue allowed us to identify two themes that are central for the management of online communities: (1) governance and (2) symbolic value creation. As argued in this Special Issue, these themes have implications for how we think about online communities and how firms can go about managing them.

The papers in this collection contribute to an ongoing debate on open, distributed, cumulative and collaborative innovation processes by highlighting the role of individual users embedded in online communities. The papers highlight a number of intriguing new perspectives and introduce new empirical settings, with implications for research and managerial practice. After having read them and the concluding research note, answers to some of the important questions about the nature of online communities emerge, but many new questions proliferate.

This emphasizes that studies on online communities for innovation may propose important improvement to conventional theories developed for other empirical settings. It appears that researchers interested in online communities have been moving into this area because of an interest in the phenomenon itself. Scholars from different backgrounds and using different theoretical perspectives have all written contributions to how online communities work. As a consequence, the body of knowledge is diverse and points in various directions. Apart from only being fascinated by the research settings—we as a community of scholars—could do a better job in building theory and think more carefully about the theoretical implications of our research. This was also discussed at the conference leading up to this Special Issue by two keynote speakers. In the first keynote at the track, Joel West discussed how different theoretical streams fit together. In the second keynote, Karim Lakhani talked about how researchers within the community define the core concepts. Indeed, as argued by West and Lakhani in the final piece of this Special Issue, several fundamental core concepts remain ambiguously defined. In order to build a more coherent body of literature, scholars pursuing research in this area therefore need to think carefully about these core concepts.

The existence of online communities, we argue, has many implications for how we think about innovation processes, and we hope that researchers will continue to address the issue, extending research and our understanding in this area.

Acknowledgements

In the autumn and winter of 2006 a call for papers was issued inviting researchers to submit papers to the track "Managing Open Innovation Through Online Communities" at the European Academy of Management conference (EURAM) in Paris, May 2007. During a few wonderful days in Paris, a group of highly qualified scholars was brought together for a lively discussion. Given the very constructive debate, it is believed that all the authors left the track with lots of ideas and suggestions for how their papers might be improved. Participants in the track were invited to submit their papers for publication in this Special Issue. After another round of peer review, five papers were selected for inclusion. Lots of help and guidance were received on the way. First, the authors who submitted manuscripts for the EURAM conference track 2007 and for review for this Special Issue are thanked. Second, the guest editors are grateful to Mark Lorenzen for inviting them to edit this Special Issue and for assistance in bringing it to print. It has provided great opportunities to develop an active community of researchers working on online communities and their relationship to open innovation. They are also very grateful to the EURAM conference and the more than 25 Special Issue reviewers whose constructive comments significantly improved the quality of the papers. Finally, the guest editors acknowledge Taylor & Francis publishers and the Innovation and Entrepreneurship Group, Tanaka Business School, Imperial College London for financial support.

References

Chesbrough, H. (2003) *Open Innovation: The New Imperative for Creating and Profiting from Technology* (Boston: Harvard Business School Press).

Chesbrough, H. (2006) *Open Business Models: How to Thrive in the New Innovation Landscape* (Boston: Harvard Business School Press).

Dahlander, L. and Magnusson, M. G. (2005) Relationships between open source software companies and communities: observations from Nordic firms, *Research Policy*, 34(4), pp. 481–493.

Dahlander, L. and Wallin, M. W. (2006) A man on the inside: unlocking communities as complementary assets, *Research Policy*, 35(8), pp. 1243–1259.

David, P. A. and Rullani, F. (forthcoming) Dynamics of innovation in an "open source" collaboration environment: lurking, laboring and launching FLOSS projects on SourceForge, *Industrial and Corporate Change*.

Fleming, L. and Waguespack, D. M. (2007) Brokerage, boundary spanning, and leadership in open innovation communities, *Organization Science*, 18, pp. 165–180.

Henkel, J. (2006) Selective revealing in open innovation processes: the case of embedded Linux, *Research Policy*, 35(7), pp. 953–969.

Jeppesen, L. B. and Frederiksen, L. (2006) Why do users contribute to firm-hosted user communities? The case of computer-controlled music instruments, *Organization Science*, 17(1), pp. 45–63.

Kogut, B. and Metiu, A. (2001) Open-source software development and distributed innovation, *Oxford Review of Economic Policy*, 14(2), pp. 248–264.

Laursen, K. and Salter, A. J. (2006) Open for innovation: the role of openness in explaining innovation performance among UK manufacturing firms, *Strategic Management Journal*, 27, pp. 131–150.

Lee, G. K. and Cole, R. E. (2003) From a firm-based to a community-based model of knowledge creation: the case of the Linux kernel development, *Organization Science*, 14(6), pp. 633–649.

Markus, M. L. (2007) The governance of free/open source software projects: monolithic, multidimensional, or configurational?, *Journal of Management Governance*, 11, pp. 151–163.

Muniz, A. M. and O'Guinn, T. C. (2001) Brand community, *Journal of Consumer Research*, 27, pp. 412–432.

O'Mahony, S. (2003) Guarding the commons: how community managed software projects protect their work, *Research Policy*, 32, pp. 1179–1198.

O'Mahony, S. and Ferraro, F. (2007) Governance in production communities, *Academy of Management Journal*, 50(5), pp. 1079–1106.

Ravasi, D. and Rindova, V. P. (2007) Symbolic value creation. Technology, Innovation and Institutions Working Paper Series, TII-16, pp. 1–32.

Rindova, V. P. and Petkova, A. P. (2007) When is a new thing a good thing? Technological change, product form design, and perceptions of value for product innovations, *Organization Science*, 18, pp. 217–232.

Shah, S. (2006) Motivation, governance, and the viability of hybrid forms in open source software development, *Management Science*, 52(7), pp. 1000–1014.

von Zedtwitz, M. and Gassmann, O. (2002) Market versus technology driven in R&D Internationalisation: four different patterns of managing research and development, *Research Policy*, 31(4), pp. 569–588.

Of Hackers and Hairdressers: Modularity and the Organizational Economics of Open-source Collaboration

RICHARD N. LANGLOIS & GIAMPAOLO GARZARELLI

ABSTRACT Using the idea of modularity, we study the general phenomenon of open-source collaboration, which includes such things as collective invention and open science in addition to open-source software production. We argue that open-source collaboration coordinates the division of labor through the exchange of effort rather than of products: suppliers of effort self-identify in the same way as suppliers of products in a market rather than accepting assignments like employees in a firm. We suggest that open-source software (and other) projects are neither bazaars nor cathedrals, but hybrids manifesting both voluntary production and conscious planning.

Introduction

We tend to think of decentralized intellectual collaboration as a phenomenon of open-source software, the Internet and the New Economy. But consider the plight of a certain Monsieur Prony, as recounted by Charles Babbage (1835: 243). Gaspard Riche de Prony (1755–1839) was a French civil engineer whom the Revolutionary government of 1790 had charged with an unenviable assignment: construct the largest and most accurate set of trigonometric and logarithmic tables ever produced. One day, while pondering this seemingly impossible task, Prony wandered into a bookseller's shop and absent-mindedly thumbed through a copy of Adam Smith's *Wealth of Nations* (1976). Suddenly it struck him. He could enlist the division of labor to construct the tables—in effect, he could manufacture logarithms like pins. Driven by this insight, Prony set up a collaborative project along the following lines. He would begin by enlisting four or five of the most eminent mathematicians in France to devise formulas well suited for numerical calculation. The results would then pass to a small team of run-of-the-mill mathematicians who would turn the formulas into simple algorithms. The actual calculations would be performed by a large team of 60–80, most of whom knew no math beyond simple addition and subtraction. Indeed, in the event the calculators came largely from the ranks of unemployed hairdressers, a group who had lost their once elaborately coiffed clients to the same Revolutionary taste for austerity and reason that had inspired Prony's commission. "By 1794, 700 results were being produced each day" (Grattan-Guinness, 1990: 180).

Obviously, this example differs from present-day open-source software projects in a number of respects. For one thing, Prony's calculators were not volunteers, and were presumably paid either as employees or on a per-calculation basis.[1] Moreover, the entire process was "fordist" rather than collegial: the design was top down, with no communication among the calculators or feedback from them to the mathematicians above. And the calculators were not creative programmers but deskilled automatons whom Babbage had every hope of replacing with his planned difference engine.[2] Nonetheless, like present-day open-source efforts, the Prony Project was an attempt to share out a complex creative task. Both are examples of what we shall call here the intellectual division of labor.[3]

This article seeks to understand the phenomenon of open-source collaboration by placing it within this larger context of the intellectual division of labor. In fact, although "open-source" has its original technical meanings in software design and law, we intend to more precisely understand the more general open-source production mode. Examples include the phenomenon of "collective invention" (Osterloh and Rota, 2004) within industrial communities (Allen, 1983) and the professions (von Hippel, 1987; Savage, 1994); the business of journal editing and refereeing familiar to academics (Bergstrom, 2001); online open bibliographic databases, such as Research Papers in Economics (RePEc) (Krichel and Zimmermann, 2005); cases of "open-source" collaboration for literary and hobbyist ends rather than for software production (like the online encyclopedia Wikipedia and the photography site photo.net); and even the modern practice of "open science" (David, 1998).

Our primary focus is on how open-source collaboration *coordinates* the intellectual division of labor. In Prony's case, the answer was top-down design and control. But the hallmark of many present-day open-source efforts is bottom-up coordination. The Prony

---

[1] Unfortunately, no information seems to have survived about the actual organization of the work (Grattan-Guinness, 1990: 179).

[2] There is at least one recent analogue to the Prony Project: NASA's Clickworkers study, which enlisted volunteers to engage in well-specified and relatively menial astrometric tasks. See: http://clickworkers.arc.nasa.gov/top. Of course, Babbage's dream of handing off the most menial calculations to computers has long since come true, and today Google will help you donate free time on your Internet-connected computer to solving a small piece of a large scientific problem. See: http://toolbar.google.com/dc/offerdc.html.

[3] What Babbage (1835: *passim* Part 3) called the "mental division of labor".

Project was fordist; but the paradigmatic open-source effort of today mostly seems to be a horizontal organization where coordination is more often unplanned rather than planned.

To answer the question of how open-source collaborations coordinate the intellectual division we turn to what we may loosely call the modularity theory of the firm (Narduzzo and Rossi, 2005; Baldwin and Clark, 2006; Langlois, 2006), since, as we will argue, open-source collaboration ultimately relies on the institutions of modularity. Our answer hinges on a tradeoff between the benefits (and costs) of modularity and the benefits (and costs) of its opposite, integrality. This tradeoff determines the extent to which central coordination is preferable to decentralized coordination. In this respect we are attempting to make more precise an idea that is already implicit in many discussions of the open-source phenomenon in software: we try to move a step beyond the many discussions of open-source production that still reason according to the *cathedral* and the *bazaar* polar ideal types (Raymond, 2001) by considering the more nuanced characteristics of actual organizational forms.[4] In effect, we argue that the organizational menu is not polar but multidimensional, where forms simultaneously exhibit different degrees of modularity (bazaar) and integrality (cathedral). More specifically, we find that open-source collaboration is a hybrid organizational form that, because of an innate modularity–integrality tradeoff, permits the exchange of effort rather than the exchange of products, and it does so under a regime in which suppliers of effort self-identify like suppliers of products in a market rather than accepting assignments like employees in a firm. We point out, moreover, that it is the nature and intensity of demand—especially the extent to which demand is quality or time sensitive—that can shift the margin between modularity and integrality. And we further suggest that the existence of this tradeoff between modularity and integrality can constitute an incentive and "focal point" for technological or organizational change (Rosenberg, 1976)—change that sometimes, and perhaps typically, shifts the margin in favor of modularity.

Modularity and Integrality

In his famous account of the development of the operating system for the IBM 360 series of computers, Frederick Brooks (1975) paints a depressing portrait of the intellectual division of labor. Like Prony, Brooks parceled out the tasks of mental labor to a large number of workers. Unlike Prony, however, Brooks found himself faced with a daunting problem of coordination. An operating system is an immensely complicated tangle of interconnections, and every piece potentially depends on every other piece. Brooks took this to mean that every programmer ought to know what every other programmer is doing, and concluded that coordination costs should rise as the square of the number of workers—an idea now sometimes called Brooks's law.[5] The implication, he reasoned, is that coordination costs quickly vitiate any benefits from the intellectual division of labor.

At about the same time, however, David Parnas (1972) and other researchers were looking at software systems in a different light. Solving the problem of interdependency, they argued, is not a matter of maximizing communication among the parts but rather of

---

[4] For an early statement that the cathedral and the bazaar referred to ideal types rather than to concrete organizational forms, see Eunice (1998).

[5] As open-source guru Eric Raymond observes, Brooks's dire conclusion rested on the implicit "assumption that the communication structure of [a] project is necessarily a complete graph" (Raymond, 2001: 35).

minimizing communication. In this approach, which laid the foundations for object-oriented programming, one attempts to design systems in which not only do the parts not need to communicate extensively with one another but are actually *forbidden* from communicating with one another. The basic idea is that "system details that are likely to change independently should be the secrets of separate modules; the only assumptions that should appear in the interfaces between modules are those that are considered unlikely to change" (Parnas *et al.*, 1985: 260). This is the notion of *information hiding* or *encapsulation*.

Software design reveals the logic of modularity with particular clarity. But the basic ideas were long ago articulated by Herbert Simon (1962) in a more general context. Simon contrasted systems that are *non-decomposable* with systems that are (or are nearly) *decomposable*. Brooks's conception of the 360 operating system created a non-decomposable system: every part communicates with virtually every other part.[6]

Consider Figure 1, where an entry of x in location a_{ij} means that element a_i communicates with element a_j. Matrix 1 is a fully non-decomposable system: every element communicates with every other element. By contrast, Matrix 2 is a relatively decomposable system. Communication is encapsulated within clusters of elements that do not communicate with elements "far away". Notice, however, that, although a nearly decomposable system like Matrix 2 clearly solves the problem of coordination, it does so by throwing the baby out with the bath water. Matrix 2 is a congeries of autarkic clusters that eliminates the costs of cooperation by the expedient of eliminating cooperation.

The term *modular system* takes on many meanings in the literature; but one important candidate definition, which we adopt here, is that a modular system is a nearly decomposable system that preserves the possibility of cooperation by adopting a common interface. The common interface enables, but also governs and disciplines, the communication among subsystems.[7] In terms of Figure 1, an interface would be a set of elements that communicates with most or all the other elements. In Matrix 3, element a_1 is the common interface: a_1 communicates with all the a_{ij} and all the a_{ij} communicate with a_1.[8] In other respects, however, Matrix 3 remains sparse off the diagonal.

Let us refer to a common interface as *lean* if it enables communication among the subsystems without creating a non-decomposable system, that is, if it enables communication without filling up the off-diagonal. As we will see, an interface may become standardized; it may also be "open" as against "closed". But it is the leanness of the interface, not its standardization or openness, that makes a system modular. Baldwin and Clark (2000) suggest thinking about modularity in terms of a partitioning of information into *visible design rules* and *hidden design parameters*. The visible design rules (or visible information) consist of three parts. (1) An *architecture* specifies what modules will be part of the system and what their functions will be. (2) *Interfaces* describe in detail how the modules will interact, including how they fit together and communicate. And (3) *standards* test a

[6] As Baldwin and Clark (2006) rightly insist, the notion of "communication" should involve more than information, and can include the transmission of energy and materials as well as symbols.

[7] This depiction of an interface is only meant to be suggestive. For one thing, cooperation may not require that communication be completely bidirectional in all cases, that is, it may not require that row 1 and column 1 be fully populated. Moreover, many different kinds of interface configurations are possible. On this point see Ulrich (1995).

[8] Think of a_1 as M. Prony's assistant, who must have had to scurry among the calculators, each one working in isolation, to gather up the results and arrange them for typesetting.

	a_1	a_2	a_3	a_4	a_5	a_6	a_7
a_1	X	X	X	X	X	X	X
a_2	X	X	X	X	X	X	X
a_3	X	X	X	X	X	X	X
a_4	X	X	X	X	X	X	X
a_5	X	X	X	X	X	X	X
a_6	X	X	X	X	X	X	X
a_7	X	X	X	X	X	X	X

Matrix 1. A non-decomposable system.

	a_1	a_2	a_3	a_4	a_5	a_6	a_7
a_1	X	X					
a_2	X	X					
a_3			X	X			
a_4			X	X			
a_5					X	X	
a_6					X	X	
a_7							X

Matrix 2. A nearly decomposable system

	a_1	a_2	a_3	a_4	a_5	a_6	a_7
a_1	X	X	X	X	X	X	X
a_2	X	X	X				
a_3	X	X	X				
a_4	X			X	X		
a_5	X			X	X		
a_6	X					X	X
a_7	X					X	X

Matrix 3. A modular system with common interface.

Figure 1. Different types of systems

module's conformity to design rules and measure the module's performance relative to other modules. Notice that "standards" in this sense doesn't necessarily imply that the architecture or interfaces are standard in the sense of being publicly shared or common to many similar artifacts or systems. A personal computer, for example, could be modular in Baldwin and Clark's sense but still be a unique design incompatible with the architecture and interfaces of other computers.[9]

All in all, then, modularity seems a powerful and elegant solution to the problem of coordinating the intellectual division of labor.[10] No less a figure than Linus Torvalds has testified to the value of modularity in the arena of open-source software.

> With the Linux kernel it became clear very quickly that we want to have a system which is as modular as possible. The open-source development model really requires this, because otherwise you can't easily have people working in parallel. It's too painful when you have people working on the same part of the kernel and they clash.
>
> Without modularity I would have to check every file that changed, which would be a lot, to make sure nothing was changed that would effect anything else. With modularity, when someone sends me patches to do a new filesystem and I don't necessarily trust the patches per se, I can still trust the fact that if nobody's using this filesystem, it's not going to impact anything else.... The key is to keep people from stepping on each other's toes. (Torvalds, 1999: 108)

Nonetheless, we argue, there can be costs to modularity as well as benefits, and there can be benefits to non-decomposability—or *integrality*, as we will call it—as well as costs. Indeed, the two are opposite sides of the same coin. What modularity does well integrality does poorly, and what integrality does well modularity does poorly. The costs of the one are essentially the foregone benefits of the other.

The Tradeoff between Modularity and Integrality

The first kind of benefit from modularity has already occupied us extensively: the ability of a modular system to obviate widespread communication among the modules (or their creators) and to limit unpredictable interactions. In effect, the process of modularization unburdens the system's elements of the task of coordination by handing that function off to the visible design rules. Coordination is imbedded or institutionalized in the structure of the system, which means that it doesn't have to be manufactured on the spot by the participants.

A second source of benefits derives from what Garud and Kumaraswamy (1995) call *economies of substitution*. We can think of these economies of substitution as a species of what economists call economies of scope. Economies of scope exist when it is cheaper to make a given product if you are already making similar products than if you were to start from scratch. This is possible to the extent that you can reuse existing fixed investments (including knowledge) instead of reinventing the wheel. Economies of scope are normally discussed as a property of the "production function" of a firm. But as Langlois and Robertson (1995: 5) argue, there can be *external* economies of scope in an open modular

---

[9] To put it another way, the term "standards" is sometimes used to mean a set of normative criteria and sometimes to imply replication. Although ultimately related, these are two quite different things. On the various meanings of the term, see David (1987).

[10] Garud *et al.* (2003) collects together many of the foundational articles on modular systems.

system, since the visible design rules constitute a shared fixed investment that everyone can reuse in creating products through substituting and recombining modules. To the extent that the interfaces governing the modular system are sufficiently standardized, it may be possible to upgrade a system by piecemeal substitution of improved modules without having to redesign the entire system. In large open-source software projects, for instance, the "approach of substituting individual components is the norm".[11] It may also be possible to optimize a system by choosing the best available modules or to customize a system to one's tastes or needs by selecting only some modules and not others[12] (Langlois and Robertson, 1992).

A third, and perhaps most important, benefit of modularity is that it militates in favor of specialization and the effective use of local knowledge. If we do not subdivide tasks, everyone must do everything, which means that everyone must know how to do everything. But, as Babbage understood, if we do subdivide tasks, we can assign workers according to comparative advantage. Why pay a mathematician for those parts of the work that an (unemployed) hairdresser could do? More significantly, however, a modular system can do more than use a given allocation of local knowledge effectively—it can potentially tap into a vast supply of local knowledge (Langlois and Robertson, 1992). This has not been lost on open-source developers, who often wax poetic on the ability of the open-source model to tap into a larger "collective intelligence".[13] Raymond even installs a version of this idea as "Linus's law": "Given enough eyeballs, all bugs are shallow." That is to say: "Given a large enough beta-tester and co-developer base, almost every problem will be characterized quickly and the fix obvious to someone" (Raymond, 2001: 30). A modular system increases the potential number of collaborators; and the larger the number of collaborators working independently, the more the system benefits from rapid trial-and-error learning[14] (Nelson and Winter, 1977; Langlois and Robertson, 1992; Baldwin and Clark, 2000).

Notice here that, unlike the first two kinds of benefits from modularity—institutionalized coordination and economies of substitution—the benefits of tapping into "collective intelligence" depend not only on the technological characteristics of the system itself but also on the way the intellectual division of labor is organized. M. Prony could have set his hairdressers to work under one roof or he could have "put out" the calculations to them in their homes; he could have paid them by the calculation or by the hour. Presumably one set of alternatives would have worked best, and the economics of organization might shed light on which set that would be. But in all cases, the modularization into simple calculations obviated communication among the calculators and with their handlers. Similarly, IBM could have—and probably did—benefit from economies of substitution in the hardware aspects of

[11] Martin Michlmayr, former Debian GNU/Linux Project Leader, personal communication, 31 December 2004. On external scope economies in the case of software, see Baetjer (1998: 99).

[12] Kuan (2001) and Bessen (2006) see the benefits of open-source software production in terms of its ability to fine-tune the product to user needs by making users part of the production process.

[13] For example: "The power of the commercial open-source business model is that you're tapping into the collective intelligence of your community" (John Roberts of SugarCRM, quote in LaMonica, 2004).

[14] As Raymond further elaborates, "while coding remains an essentially solitary activity, the really great hacks come from harnessing the attention and brainpower of entire communities. The developer who uses only his or her own brain in a closed project is going to fall behind the developer who knows how to create an open, evolutionary context in which feedback exploring the design space, code contributions, bug-spotting, and other improvements come from hundreds (perhaps thousands) of people" (Raymond, 2001: 50–51).

the 360 series of computers, even though the system was emphatically closed to outside collaborators, the so-called plug-compatible vendors.[15] But for a modular system to take advantage of extended localized knowledge, the organization of the intellectual division of labor is no longer immaterial. In order to tap into "collective knowledge", the system's interface must be not only lean but also relatively standardized and open.

The economics of networks has taught us that, despite the occasional subtlety, standardization is the easy part. If interfaces are sufficiently lean and sufficiently open, there is a tendency for one of them to emerge as a dominant standard (Shapiro and Varian, 1998). Openness is the more interesting issue. As the community of open-source software developers clearly understands, openness does not mean only unfettered access to *knowledge* of the visible design rules of the system, though that may be a necessary condition. Rather, openness is about the *right* to take advantage of those design rules.[16] More broadly, the degree of openness of a modular system is bound up with the overall assignment of *decision rights* within the intellectual division of labor.[17] This is an organizational issue and—what may be the same thing—a *constitutional* issue that is taken up in some detail in the next section.

Let us now consider the other side of the coin: the costs of modularity. The first of these is the (fixed) cost of establishing the visible design rules (Baldwin and Clark, 2000). A (nearly) decomposable system may solve coordination problems in an elegant way, but designing such a system may take a considerable amount of time and effort.[18] There may also be costs to communicating the design rules to participants and securing agreement on them.

Another cost is that, at least in principle, it may not be possible to fine-tune a modular system as tightly as an integral one. For many kinds of software, this may no longer be an important issue in the face of Moore's law.[19] But for other kinds of systems, there may be important performance losses from building a system out of modules. Automobiles, for example, may have an inherent "integrality" that prevents automakers from taking advantage of modularity to the same degree as, say, makers of personal computers (Helper and MacDuffie, 2002). One can't swap engines like one swaps hard drives, since a different engine can change the balance, stability and handling of the entire vehicle. Clayton Christensen and his collaborators (Christensen *et al.*, 2002) have argued that integral

---

[15] Who famously sued IBM under antitrust laws in an attempt to open access to the system. *United States v. International Business Machines Corporation*, 1956 Trade Cas. #68, 245 (S.D.N.Y. 1956).

[16] The question of rights in open-source software is a much-discussed issue, revolving around the various kinds of software licenses possible or in actual use. It would take us too far afield to enter this complex area. But for some distinctions see Wheeler (2007a, b; n.d.).

[17] At this point we have in mind the allocation of decision rights within the intellectual division of labor—for example, the allocation of decision rights to programmers in the process of designing a piece of software. But the same logic applies at the level of the system—for example, the software—itself. In that case, we can say that the visible design rules themselves are a kind of "constitution" that determines the rights of action that the elements of the system enjoy. And here there is a strong analogy between the idea of encapsulation in software design and private property rights in legal systems (Miller and Drexler, 1988; Langlois, 2006). In both cases, the operative principle is the creation of a protected sphere of activity permitting autonomy of action. The two levels overlap to the extent that a particular encapsulated subsystem is the property of a particular individual or group within the intellectual division of labor.

[18] Garud and Kumaraswamy (1995) cite evidence that the cost of designing a reusable modular software object may be as much as 10 times the cost of designing software intended to be used only once.

[19] After Gordon Moore, Intel co-founder, who claimed that the number of functions that can be crammed on a chip doubles every 12 months (Moore, 1965).

designs, which can take advantage of systemic fine-tuning, have an advantage whenever users demand higher performance than existing technology can satisfy. As the fine-tuned system continues to improve in performance, however, it will eventually surpass typical user needs. At that point, these authors argue, production costs move to the fore, and the integral system (and the integrated organization that designed it) will give way to a network of producers taking advantage of the benefits of modularity discussed earlier.

At the same time, however, one might also add that sometimes a modular system can improve in performance even faster than a fine-tuned system. To the extent that such a system benefits from "collective intelligence" and rapid-trial-and-error learning, the improvement in the parts can dominate any benefits from fine-tuning. Personal computers are again a case in point. PCs have come to outperform first mainframes, then minicomputers, then RISC workstations, all of which, in their day, made their money as fine-tuned non-modular systems (at least relative to PCs). Again, the extent to which modular innovation can outperform fine-tuning may depend on the degree of inherent integrality in the system.

A third, closely related, cost of modularity (benefit of integrality) is the tendency of modular systems to become "locked in" to a particular system decomposition. At least to the extent that knowledge gained creating one modularization of the system cannot be reused in generating a new decomposition, it is a relatively costly matter to engage in systemic change of a modular system since each change requires the payment anew of the fixed cost of setting up visible design rules. If in addition an interface has become a standard, the problems of lock-in are compounded in the way popularized by Paul David (1985), since in that case many people would have simultaneously to pay the fixed cost of change. A modular system is good at modular or *autonomous* innovation, that is, innovation affecting the hidden design parameters of a given modularization but not affecting the visible design rules. But a modular system is bad at *systemic* innovation, that is, innovation that requires simultaneous change in the hidden design parameters and the visible design rules—simultaneous change in the parts and in the way the parts are hooked together.[20]

The benefits of an integral system in systemic change are related to the benefits of fine-tuning to which Christensen *et al.* point. Fine-tuning is after all systemic change to improve performance. Thus integral systems may have advantages not only when users demand high performance in a technical sense but also when they need performance in the form of change and adaptability. This latter may also be a function of how *quickly* the user needs the system to perform; the front-end costs of a modular system may take the form of time costs—the output forgone while waiting for the modularization to crystallize or the visible design rules to get worked out. If a modularization is already in place, of course, the system can adapt and respond quickly by simply plugging in new modules to suit user needs.[21] But if there is not yet a modularization, or if the user needs a level of performance greater than

[20] The terms autonomous and systemic are from Teece (1986). There is a third possibility, what Henderson and Clark (1990) call *architectural* innovation. Here the modules remain intact, but innovation takes place in the way the modules are hooked together. (For a paradigmatic example of this kind of innovation, visit Legoland.) The possibility of architectural innovation underlies the benefits of economies of substitution discussed earlier.

[21] It is for this reason that Figure 3 grades modular systems as "B" rather than "C" on their ability to fine-tune performance or adapt systemically. Modular systems can in fact fine-tune and adapt systemically to some degree by taking advantage of economies of substitution.

can be achieved even with the best possible assortment of available modules, then an integral system may do better.

Spontaneity and Design in Software Production

Effort and Division of Labor

Open-source software production is an organizational form that to a large extent relies on the institutions of modularity driven by exactly the supply side and demand side factors to which Langlois and Robertson (1992) long ago pointed. On the demand side are idiosyncratic user tastes and requirements that call for both high quality and fine degree of customization. On the supply side are the benefits of specialization by comparative advantage and the external economies of a large and diverse talent pool. Consider this description of the Debian Project.[22]

> The Debian design process is open to ensure that the system is of the highest quality and that it reflects the needs of the user community. By involving others with a wide range of abilities and backgrounds, Debian is able to be developed in a modular fashion. Its components are of high quality because those with expertise in a certain area are given the opportunity to construct or maintain the individual components of Debian involving that area. Involving others also ensures that valuable suggestions for improvement can be incorporated into the distribution during its development; thus, a distribution is created based on the needs and wants of the users rather than the needs and wants of the constructor. It is very difficult for one individual or small group to anticipate these needs and wants in advance without direct input from others.[23]

But what is the more specific organizational nature of open-source software development efforts?

To answer this question, we present a taxonomy of production models by reasoning along the following lines. Markets are about the exchange of products or outputs; such exchange is coordinated spontaneously, in the sense that relative prices rather than fiat direct resources. A firm stands in contrast to both of these aspects of markets: it replaces contracts for products with employment contracts, effectively substituting a factor market for a product market (Cheung, 1983); at the same time, it replaces spontaneous coordination with some kind of central design or direction. Notice that this leaves two unexamined alternatives: product markets governed by central direction and factor markets coordinated spontaneously. Inside contracting and outsourcing are examples of the former. Open-source collaboration is an example of the latter. In the final analysis we find that open-source collaboration manifests both spontaneity (self-selection in terms of which input to contribute) *and* design (conscious direction).

---

[22] Debian is an association of volunteers who work on an operating system called Debian GNU/Linux. GNU is a recursive acronym that stands for "GNU's Not UNIX"; it is pronounced "guh-NEW" (http://www.gnu.org/). Linux refers to the *kernel*—the central functions—of the operating system; GNU/Linux is the complete operating system, including the Linux kernel along with other software components. A *distribution* is simply a (usually complete) packaging of the Linux kernel with other software needed to complete the operating system. Debian GNU/Linux is a GNU/Linux distribution that is produced entirely by volunteers.

[23] http://www.debian.org/doc/manuals/project-history/ap-manifesto.en.html

As Jensen and Meckling (1992: 251) point out, economic organization must solve two different kinds of problems: "the rights assignment problem (determining who should exercise a decision right), and the control or agency problem (how to ensure that self-interested decision agents exercise their rights in a way that contributes to the organizational objective)". All other things equal, efficiency demands that the appropriate knowledge finds its way into the hands of those making decisions. There are basically two ways to ensure such a "collocation" of specific knowledge and decision-making: "One is by moving the knowledge to those with the decision rights; the other is by moving the decision rights to those with the knowledge" (Jensen and Meckling, 1992: 253). These two choices—as well as possible variants and hybrids—are "constitutions" that set out the assignment of decision rights. Such assignments can take place within firms or within wider networks of independent collaborators.[24]

In the case of the Prony Project, and of fordist production generally, decision rights remain centralized. This is because there is very little knowledge that needs to be transmitted; tasks have been made exceedingly simple, and the important knowledge—that involving design—is already at the center. The agency problem can be addressed either through investments in monitoring or by aligning incentives using a suitable piece rate. Even when the subdivided tasks are far more complicated—and require far more skill and creativity—it is still possible to organize the intellectual division of labor in more or less the same way. In this model—which we call *corporate*—the ultimate decision rights remain centralized, even as many de facto decision rights are parceled out to employees at various levels of the hierarchy. Clearly, such an arrangement complicates the agency problem, since keeping everyone on the same page is no longer a simple matter of monitoring or incentive alignment in a narrow pecuniary sense.

Many would argue that, even within the corporate context, effective management of high-human-capital projects requires recourse to more "participatory" or collaborative models (Minkler, 1993). Does this mean that there is really no difference between the corporate model and more decentralized ones? The answer is no, for two reasons. First, as we have seen, even a large organization is bounded in the capabilities on which it can draw, and this limitation may be important in many cases. Second, the location of the ultimate decision rights matters. For any division of intellectual labor we choose, behavior and performance will be different if we assign decision rights to some central authority rather than to the individual collaborators.[25]

The opposite of a corporate model would be a fully decentralized one in which the collaborators retain the ultimate decision rights. But just as the central holder of decision rights in a corporation must in practice cede de facto decision rights to others, so in a decentralized system the collaborators must give up some pieces of their rights in order to

---

[24] On the idea that firms are constitutional systems, see Gifford (1991).

[25] Oliver Williamson (1985: 136) traces this effect to the phenomenon of "selective intervention", the tendency of the central holder of decision rights to meddle in the decisions of the collaborators. Henry Hansmann (1996) reminds us that those who possess de facto decision rights will be constrained in their exercise of those rights if they lack the ultimate or "formal" decision rights (which Hansmann equates with ownership). The modern corporation, he points out, is precisely an example in which it pays to assign the formal ownership rights to parties (the stockholders) who may actually be in a poor position to exercise effective control—in part because such an allocation keeps the rights out of the hands of other parties (notably the managers) who would abuse them.

collaborate. In a classic market narrowly understood, the collaborators do this through contract. In return for compensation from you, I choose to exercise my right to make shoes by producing the kelly green golf shoes you have contracted for.[26] In the limit, however, I may not even know who "you" are, and I exercise my decision rights in the direction of kelly green golf shoes in the hope that you, or someone like you, comes along. A classic market of this sort is an example of what we call the *spontaneous* model—spontaneous in the sense that the division of labor itself emerges (in the limit) from the choices of the collaborators rather than from a central designer.

This is a perspective on the firm/market dichotomy somewhat different from what one usually encounters in the economics of organization. More typically, one hears the following kind of story: markets are good at exchanging *products* for compensation, whereas firms are good at exchanging *effort* for compensation. The economics of organization can be understood from this perspective as a set of stories about why it is often costly to cooperate by trading products and often necessary to cooperate by trading effort.[27] Ever since Coase (1937), it has been more or less taken for granted that the only way to trade effort is through an employment contract: I pay for your time and the right to direct your effort within agreed limits (Simon, 1951). In other words, the only way to trade effort is by setting up a firm. Perhaps the most intriguing aspect of the open-source model is that it flies in the face of this assumption: under the right circumstances, it is possible to cooperate spontaneously on the effort margin, not just the product margin.[28] Rather than giving up their decision rights to others, open-source collaborators combine effort "voluntarily". Voluntarily here means not that the collaborators do not receive pecuniary compensation (though that may often be true) but rather that the collaborators choose their own tasks. Assignment of individuals to tasks—and, to an extent we will explore, even the overall design of the division of labor itself—arises from these voluntary choices, in much the same way that assignment of sellers to products in a classic market arises from self-selection.

We can distinguish four possibilities. Consider Figure 2. Along the vertical dimension is the issue of design: is assignment to task (and maybe even the division of labor itself) generated through a centralized process or does it arise from the self-identification of collaborators with tasks? Along the horizontal dimension is the problem of information and agency: are we talking about products cleanly measured and priced or are we taking about exchanges of effort that involve costs of measurement and agency? In the upper left-hand box, the division of labor is centrally designed, but the products of that labor are easily measured and priced. This is the world of inside or outside contracting. It is the Prony Project, as well as the outsourcing of intellectual activities like call centers, back-office work or the reading of X-rays. In the lower left-hand box, the division of labor remains centrally designed, but the cost of measuring and pricing transactions makes it cheaper to purchase

---

[26] Apologies to Deirdre McCloskey (1995).

[27] This is not a denial that effort is ultimately a product. It is merely a claim that effort is a particular kind of product, one whose properties make it costly to trade using only the relatively rudimentary institutional support that an anonymous spot market offers.

[28] As pointed out earlier, of course, this model actually antedates software. Notably, it has been the normal mode of organization in the professions (Savage, 1994), including those that produce science (David, 1998). But only with the prominence of open-source software development has the phenomenon begun to gain the attention of the economics of organization.

	Don't self-identify	Self-identify
Products	Inside contracting Outsourcing	Classic market
Effort	Classic firm	Voluntary production

Figure 2. Governance possibilities

the effort of collaborators directly. This is the classic firm. In the upper right-hand box, participants self-select their contributions; but measurement and pricing costs are not prohibitive, and those contributions take the form of products offered on spec. This is the classic market. Finally, in the lower right-hand box, participants self-select their contributions; but those contributions come directly in the form of effort rather than of effort embodied in a product. That is, we do not have, for instance, a spot market in day labor where day laborers don't choose what they work on; rather, we have a division of labor where programmers do sometimes produce specific products in the end, but in the context of spontaneously adding effort to a larger product not making a product on spec. This is the model of *voluntary* production, the model that most closely resembles the bazaar ideal type.

Rules and Division of Labor

This two-dimensional schema has advantages, we argue, over the tripartite distinction Benkler (2002) offers among markets, hierarchies and what he calls *peer production*. Benkler argues that a perceptible trend toward the increased importance of human capital in production is leading toward peer production and away from both markets and hierarchies. It may well be that, with economic growth and an expanding extent of the market, there is a general trend rightward in Figure 2, what Langlois (2003) calls the phenomenon of the Vanishing Hand. But an increased importance of human capital and greater spontaneity of production is consistent with markets as well as with decentralized collaboration through direct effort. Even apart from the likes of books, musical scores or screenplays, there are a plethora of "consulting" services—from legal representation to brain surgery—that are priced on markets.

The most extreme form of a voluntary arrangement would occur when the self-selection of the collaborators itself actually creates the division of labor. This is far from unimaginable: it is exactly what happens in "the market" in the largest sense—including the market for software in the large. It also arguably happens in the context of academic open science, where the pattern of knowledge emerges from the self-selected research choices of the participants. If we cast our gaze down to a less lofty level, however, there almost always seems to be some pre-existing structure of possible tasks from which the participants choose. If Thomas Kuhn (1970) and others are right, even scientific researchers are often—and maybe always—guided in their choice of problem by the constraints of earlier models and approaches. And at the level of any particular software project, the self-selection of workers to tasks takes place within the context of an established architecture or (at the very least) an established technological trajectory.[29]

Why is this so? Consider the experiments conducted by Kevan Davis, a British software engineer (Thompson, 2004). Davis set up a website on which visitors could vote, pixel by pixel, on the design of a type font.[30] The results were quite presentable and, at least when cleaned up a bit, looked "like a mildly punk version of Helvetica, with occasional flashes of creative weirdness, such as the jaunty serif on the foot of the letter 'J'" (Thompson, 2004). But when Davis asked people to draw a face or a television set by voting on pixels, the result was a shapeless mess. Efforts at drawing a goat looked mildly goat-like for a while, then collapsed into a jumble after 7,000 votes. The difference between a type font and a goat, of course, is that we come prepared with structural preconceptions that are far tighter in the former case than in the latter. For the font problem, there is something much closer to a pre-existing design architecture to guide individual contributions. This shouldn't be surprising in view of our discussion of modularity. Modularity enables large-scale cooperation; but it requires agreed-upon visible design rules (which, as in the case of letters, may even be tacitly known design rules).

In a typical well-run open-source project, design structure takes the form both of conscious direction and of anonymous rules. It is the first of these elements that makes the development process ultimately a "hybrid" form of organization. Conscious direction is most important in the early and relatively inchoate stages of a project—when design issues remain systemic in important respects. As Ian A. Murdock, creator of Debian, for example recalls:

> Debian's biggest organizational achievement was its emphasis on project management and infrastructure. It's one thing to have a great idea that generates interest; it's another to have the necessary infrastructure in place so that, when the masses show up to lend a hand, they can contribute.
>
> In the early days of Debian, the package system and packaging standards we put together ensured that independently developed pieces came together into a cohesive whole. Later, as the number of Debian developers swelled from dozens to hundreds to more than 1,100, a project management infrastructure took shape to handle this massive task.
>
> This, by the way, is why most open-source projects never get beyond the idea phase (just browse through any 20 random projects on SourceForge [http://sourceforge.net/] to see what I mean). No matter how great the idea, there has to be a framework in which to contribute, or critical mass can never be reached. And without critical mass, open development projects are not sustainable. (Murdock, 2003)

As a result, Debian volunteers now follow the rules contained in the Debian Social Contract,[31] the Debian Free Software Guidelines (aka, DFSG),[32] and the

---

[29] The idea of a technological trajectory is an application to technology of Kuhn's idea of a scientific paradigm (Dosi, 1982).

[30] http://www.typophile.com/

[31] http://www.debian.org/social_contract

[32] Ibid. Incidentally, from the DFSG derives the Open Source Definition.

Constitution.[33] These visible design rules assure that "each package can be dropped into the system independently without damaging or interfering with programs from other packages. By working with a set of consistent rules and with identical tools, the volunteers can and do create a truly modular system" (Murdock, 1994a). Indeed, over time, design becomes increasingly a matter of impersonal rules, sometimes explicit, sometimes tacit.

But projects almost always retain important elements of conscious direction even as they mature. Like the impersonal design rules, however, these elements of conscious direction involve design at the highest—most abstract—levels of the design hierarchy. Because it takes place at an abstract level, direction in this kind of project plays the role of a (lean) interface among the modular parts rather than of a systemic meddler in an interconnected system.[34] This direction often rests with Project Leaders and core groups of programmers. For instance:

> The Debian leadership checks to ensure that each package is assembled correctly and that the system as a whole is solid; the end result is a set of packages that, though developed and maintained by many different individuals, are as consistent and as professionally constructed as if they were developed by a single person or closely-knit group, but without the limitations imposed by centralized development. (Murdock, 1994b)

In terms of our earlier distinction between the corporate model and the spontaneous model, the need for performance and rapid adaptability would tend to militate in the direction of the corporate (Langlois, 1988). But, as pointed out, this does not mean that unsatisfied needs for performance and rapid systemic adaptation therefore call for central planning in the socialist sense. In Christensen *et al.*'s account, unmet performance needs do always call for an integrated corporate structure. But the network theorist Duncan Watts (2004) reminds us that a decentralized structure, with its ability to utilize "collective intelligence", can sometimes be marshaled even in the service of an emergency response. His example is the way the Toyota Corporation responded in 1997 when the sole plant supplying a crucial component burned to the ground, threatening to bring production of an entire model to a halt. Rather than attempting to create centrally a new plant to make the component, Toyota instead tapped the knowledge and capabilities of a large number of its divisions and outside suppliers with the intent of generating rapid trial-and-error learning.

> More than 200 companies reorganized themselves and each other to develop at least six entirely different production processes, each using different tools, different engineering approaches, and different organizational arrangements. Virtually every aspect of the recovery effort had to be designed and executed on the fly, with engineers and managers sharing their successes and failures alike across departmental boundaries, and even between firms that in normal times would be direct competitors. (Watts, 2004)

Within a week, production of the component was back to pre-fire levels.

---

[33] http://www.debian.org/devel/constitution.en.html

[34] For a more general argument that the degree of abstractness of rules is the fundamental difference between designed and spontaneous orders, see Langlois (1995).

Clearly, of course, this response was not spontaneous in our sense. It was centrally directed and coordinated to a large extent. But neither was it the standard corporate model. Rather, it is an example of what we call a *hybrid* model—one that has elements both of spontaneous, self-selected production and of central design. In essence, a hybrid model is an attempt to get around the tradeoffs summarized in Figure 3. Such hybrid models are actually far more typical of open-source software development than are genuinely voluntary or spontaneous ones. Indeed, perhaps *all* models of open-source software development are hybrid models.

	Modularity	Integrality
Communications costs	A	C
Economies of substitution	A	C
"Collective intelligence"	A	C
Set-up costs	C	A
Fine tuning	B	B
Systemic adaptation	A	B

Figure 3. Grading the alternatives

Conclusion

The organizational economics of open-source collaboration (collective invention, the professions, open science, etc.) are not simple. In light of this, we propose that an understanding of such collaboration requires reasoning according to the principles of the emerging economic theory of modularity. These principles allow us to clarify under which circumstances, and in what quantities, central design is necessary, and when a genuinely decentralized design can lead to a well-ordered and effective structure. Such clarification leads us to suggest that open-source collaboration, including the software one, is a hybrid form of organization that presents elements of both the bazaar and cathedral ideal types. More precisely, we argue that open-source collaboration is a type of intellectual division of labor that exchanges effort, where the nature of this exchange is undertaken on a voluntary basis rather than according to fiat.

Acknowledgements

Previous versions of this article have benefited from the feedback received from Davide Consoli, Greg Lee, Martin Michlmayr, audiences at DRUID 18–20 June 2006, Copenhagen and EURAM 16–19 May 2007, Paris, seminar participants at Wits on 25 April 2007 and at the University of Pretoria on 3 May 2007, and three referees of this journal. The second author moreover acknowledges the Centro di Metodologia delle Scienze Sociali at LUISS-Guido Carli, Rome, particularly Dario Antiseri and Caterina Galluccio, for hospitality from November 2006 to January 2007.

References

Allen, R. C. (1983) Collective invention, *Journal of Economic Behavior and Organization*, 4(1), pp. 1–24.

Babbage, C. (1835) *On the Economy of Machinery and Manufactures*, 4th edn (London: Charles Knight). Available at http://socserv2.socsci.mcmaster.ca/~econ/ugcm/3ll3/babbage/.

Baetjer, H., Jr. (1998) *Software as Capital. An Economic Perspective on Software Engineering* (Los Alamitos, CA: IEEE Computer Society).

Baldwin, C. Y. and Clark, K. B. (2000) *Design Rules: The Power of Modularity*, Vol. I (Cambridge, MA: MIT Press).

Baldwin, C. Y. and Clark, K. B. (2006) Where do transactions come from? A network design perspective on the theory of the firm, Working Paper, Harvard Business School, 22 May. Available at http://www.people.hbs.edu/cbaldwin/DR2/BaldwinTransactions2006.pdf.

Benkler, Y. (2002) Coase's penguin, or, Linux and the nature of the firm, *Yale Law Journal*, 112(3), pp. 369–446. Available at http://www.yale.edu/yalelj/112/BenklerWEB.pdf.

Bergstrom, T. C. (2001) Free labor for costly journals?, *Journal of Economic Perspectives*, 15(3), pp. 183–198.

Bessen, J. (2006) Open-source software: free provision of complex public goods, in: J. Bitzer & P. J. H. Schröder (Eds), *The Economics of Open Source Software Development*, Chap. 3 (Amsterdam: Elsevier). Available at http://www.researchoninnovation.org/opensrc.pdf.

Brooks, F. P. (1975) *The Mythical Man-Month. Essays on Software Engineering* (Reading, MA: Addison-Wesley).

Cheung, S. N. S. (1983) The contractual nature of the firm, *Journal of Law and Economics*, 26(1), pp. 1–21.

Christensen, C. M., Verlinden, M. and Westerman, G. (2002) Disruption, disintegration, and the dissipation of differentiability, *Industrial and Corporate Change*, 11(5), pp. 955–993.

Coase, R. H. (1937) The nature of the firm, *Economica* (N.S.), 4(16), pp. 386–405.

David, P. A. (1985) Clio and the economics of QWERTY, *American Economic Review, Papers and Proceedings*, 75(2), pp. 332–337.

David, P. A. (1987) Some new standards for the economics of standardization in the information age, in: P. Dasgupta & P. Stoneman (Eds), *Economic Policy and Technological Performance*, pp. 206–239 (Cambridge: Cambridge University Press).

David, P. A. (1998) Common agency contracting and the emergence of "open science" institutions, *American Economic Review, Papers and Proceedings*, 88(2), pp. 15–21.

Dosi, G. (1982) Technological paradigms and technological trajectories, *Research Policy*, 11(3), pp. 147–162.

Eunice, J. (1998) *Beyond the Cathedral, Beyond the Bazaar*, 11 May. Available at http://www.illuminata.com/cgi-local/pub.cgi?docid=cathedral.

Garud, R. and Kumaraswamy, A. (1995) Technological and organizational designs for realizing economies of substitution (Special issue: Technological transformation and the new competitive landscape), *Strategic Management Journal*, 16(Summer), pp. 93–109.

Garud, R., Kumaraswamy, A. and Langlois, R. N. (Eds) (2003) *Managing in the Modular Age: Architectures, Networks and Organizations* (Oxford: Blackwell).

Gifford, A., Jr. (1991) A constitutional interpretation of the firm, *Public Choice*, 68(1–3), pp. 91–106.

Grattan-Guinness, I. (1990) Work for the hairdressers: the production of de Prony's logarithmic and trigonometric tables, *IEEE Annals of the History of Computing*, 12(3), pp. 177–185.

Hansmann, H. (1996) *The Ownership of Enterprise* (Cambridge, MA: Belknap Press of Harvard University Press).

Helper, S. and MacDuffie, J. P. (2002) B2B and modes of exchange: evolutionary and transformative effects, in: B. Kogut (Ed.), *The Global Internet Economy* (Cambridge, MA: MIT Press).

Henderson, R. M. and Clark, K. B. (1990) Architectural innovation: the reconfiguration of existing product technologies and the failure of established firms, *Administrative Science Quarterly*, 35(1), pp. 9–30. Available at http://findarticles.com/p/articles/mi_m4035/is_n1_v35/ai_8305916.

Jensen, M. C. and Meckling, W. H. (1992) Specific and general knowledge, and organizational structure, in: W. Lars & H. Wijkander (Eds), *Contract Economics*, pp. 251–274 (Oxford: Basil Blackwell). Available at http://papers.ssrn.com/sol3/papers.cfm?abstract_id=6658.

Krichel, T. and Zimmermann, C. (2005) The economics of open bibliographic data provision, Working Paper 2005-01, Department of Economics, University of Connecticut, Storrs. Available at http://ideas.repec.org/p/uct/uconnp/2005-01.html.

Kuan, J. W. (2001) Open-source software as consumer integration into production, Working Paper, January. Available at http://ssrn.com/abstract=259648.

Kuhn, T. S. (1970) *The Structure of Scientific Revolutions*, 2nd edn (Chicago: The University of Chicago Press).

LaMonica, M. (2004) Breaking the rules with open-source, CNET News.com, 2 August, 4:00 AM PT. Available at http://zdnet.com.com/2100-1104_2-5290983.html.

Langlois, R. N. (1988) Economic change and the boundaries of the firm, *Journal of Institutional and Theoretical Economics*, 144(4), pp. 635–657.

Langlois, R. N. (1995) Do firms plan?, *Constitutional Political Economy*, 6(3), pp. 247–261.

Langlois, R. N. (2003) The vanishing hand: the changing dynamics of industrial capitalism, *Industrial and Corporate Change*, 12(2), pp. 351–385.

Langlois, R. N. (2006) The secret life of mundane transaction costs, *Organization Studies*, 27(9), pp. 1389–1410.

Langlois, R. N. and Robertson, P. L. (1992) Networks and innovation in a modular system: lessons from the microcomputer and stereo component industries, *Research Policy*, 21(4), pp. 297–313.

Langlois, R. N. and Robertson, P. L. (1995) *Firms, Markets, and Economic Change: A Dynamic Theory of Business Institutions* (London: Routledge).

McCloskey, D. N. (1995) Kelly green golf shoes and the intellectual range from M to N, *Eastern Economic Journal*, 21(3), pp. 411–414. Available at http://findarticles.com/p/articles/mi_qa3620/is_199507/ai_n8712423.

Miller, M. S. and Drexler, K. E. (1988) Markets and computation: agoric open systems, in: B. Huberman (Ed.), *The Ecology of Computation*, pp. 133–176 (Amsterdam: North-Holland). Available at http://www.agorics.com/agorpapers.html.

Minkler, A. P. (1993) The problem with dispersed knowledge: firms in theory and practice, *Kyklos*, 46(4), pp. 569–587.

Moore, G. (1965) Cramming more components onto integrated circuits, *Electronics*, 38, pp. 114–117. Available at http://ieeexplore.ieee.org/iel3/5/14340/00658762.pdf?arnumber=658762.

Murdock, I. (1994a) Overview of the Debian GNU/Linux system, *Linux Journal*, 6es, Article No. 15 (October).

Murdock, I. (1994b) The open development of Debian, *Linux Journal*, 3es, Article No. 7 (June–July).

Murdock, I. (2003) Debian: a brief retrospective. Available at http://www.linuxplanet.com/linuxplanet/editorials/4959/1/.

Narduzzo, A. and Rossi, A. (2005) The role of modularity in free/open source software development, in: S. Koch (Ed.), *Free/Open Software Development*, pp. 84–102 (Hershey, PA: Idea Group).

Nelson, R. R. and Winter, S. G. (1977) In search of useful theory of innovation, *Research Policy*, 6(1), pp. 36–76.

Osterloh, M. and Rota, S. G. (2004) Open-source software development—just another case of collective invention? Working Paper, University of Zurich, March. Available at http://ssrn.com/abstract=561744.

Parnas, D. L. (1972) On the criteria to be used in decomposing systems into modules, *Communications of the ACM*, 15(12), pp. 1053–1058. Available at http://www.acm.org/classics/may96/.

Parnas, D. L., Clemens, P. C. and Weiss, D. M. (1985) The modular structure of complex systems, *IEEE Transactions on Software Engineering*, 11(3), pp. 259–266. Available at http://www.cs.wm.edu/~coppit/other-papers/parnas-modular-structure.pdf.

Raymond, E. S. (2001) *The Cathedral and the Bazaar. Musings on Linux and Open-Source by an Accidental Revolutionary*, rev. edn (Sebastopol, CA: O'Reilly & Associates, Inc.). Available at http://www.catb.org/~esr/writings/cathedral-bazaar/.

Rosenberg, N. (1976) *Perspectives on Technology* (New York: Cambridge University Press).

Savage, D. A. (1994) The professions in theory and history: the case of pharmacy, *Business and Economic History*, 23(2), pp. 130–160.

Shapiro, C. and Varian, H. R. (1998) *Information Rules: A Strategic Guide to the Network Economy* (Cambridge, MA: Harvard Business School Press).

Simon, H. A. (1951) A formal theory of the employment relationship, *Econometrica*, 19(3), pp. 293–305.

Simon, H. A. (1998) The architecture of complexity: hierarchic systems, in: *Idem, The Sciences of the Artificial*, 3rd edn, 2nd printing, pp. 183–216 (Cambridge, MA: MIT Press). Originally published in 1962, *Proceedings of the American Philosophical Society*, 106(6), pp. 467–482.

Smith, A. (1976) *An Enquiry into the Nature and Causes of the Wealth of Nations*, Glasgow edn (Oxford: Clarendon Press).

Teece, D. (1986) Profiting from technological innovation: implications for integration, collaboration, licensing, and public policy, *Research Policy*, 15(6), pp. 285–305.

Thompson, C. (2004) Art mobs: can an online crowd create a poem, a novel, or a painting?, *Slate*, 21 July. Available at http://www.slate.com/id/2104087/.

Torvalds, L. (1999) The Linux edge, in: C. DiBona, S. Ockman & M. Stone (Eds), *Open-Sources: Voices from the Open-Source Revolution*, pp. 101–111 (Sebastopol, CA: O'Reilly & Associates, Inc.). Available at http://www.oreilly.com/catalog/opensources/book/toc.html.

Ulrich, K. (1995) The role of product architecture in the manufacturing firm, *Research Policy*, 24(3), pp. 419–440.

von Hippel, E. (1987) Cooperation between rivals: informal know-how trading, *Research Policy*, 16(6), pp. 291–302.

Watts, D. (2004) Decentralized intelligence: what Toyota can teach the 9/11 Commission about intelligence gathering, *Slate*, 5 August. Available at http://www.slate.com/id/2104808/.

Wheeler, D. A. (2007a) Why open-source software/free software (OSS/FS)? Look at the numbers!, 16 April. Available at http://www.dwheeler.com/oss_fs_why.html.

Wheeler, D. A. (2007b) Make your open-source software GPL-compatible. Or else, 12 June. Available at http://www.dwheeler.com/essays/gpl-compatible.html.

Wheeler, D. A. (n.d.) Open-source software/free software (OSS/FS) references. Available at http://www.dwheeler.com/oss_fs_refs.html.

Williamson, O. E. (1985) *The Economic Institutions of Capitalism* (New York: Free Press).

The Role of Participation Architecture in Growing Sponsored Open Source Communities

JOEL WEST & SIOBHÁN O'MAHONY

ABSTRACT Most research on open source software communities has focused on those that are community founded. More recently, firms have founded their own open source communities. How do sponsored open source communities differ from their autonomous counterparts? With comparative examination of 12 open source projects initiated by corporate sponsors, we identify three design parameters that together help form a participation architecture—the opportunity structure extended to potential external contributors. In exploring sponsors' community design decisions, we found that sponsored open source projects were more likely to offer transparency than they were accessibility and that this had implications for their communities' growth. We contribute theoretical constructs that offer a common basis of comparison for the future study of open source projects and illustrate how the tension between control and growth affects open source community design and creation.

Technical communities often play a crucial role in helping firms to develop and deploy new technical innovations (Rosenkopf and Tushman, 1998; Rosenkopf *et al.*, 2001; Mowery and Simcoe, 2005; Fleming and Waguespack, 2007). Technical communities provide a vehicle for the exchange of technical information that fosters the accumulation of innovations (Allen, 1983; von Hippel, 1988; Saxenian, 1994) and enables actors from different organizational forms to collaborate (van de Ven, 1993; Rosenkopf and Tushman, 1998). Technical communities, like industry associations, help firms identify mutual interests that might otherwise go undiscovered (Sabel, 1984). Empirical work suggests that firms benefit from participating in technical communities by gathering information on potential alliances, identifying opportunities for future inter-firm collaboration (Rosenkopf *et al.*, 2001) and sharing risk (Tushman and Rosenkopf, 1992; Rosenkopf and Tushman, 1994, 1998).

One type of technical community that has received a great deal of empirical attention is the open source software development community. These communities are composed of individuals who collaborate toward a common goal but do not share a common employer and are not governed by an employment hierarchy. By using both online and offline means, open source software communities collectively produce software that is freely and publicly available—creating in effect a shared public good that can be used for either public or private purposes (Lee and Cole, 2003; O'Mahony, 2003; von Hippel and von Krogh, 2003; Markus, 2007; Murray and O'Mahony, 2007).

Prior research on open source communities has emphasized autonomous and "self-managed" communities that are typically founded by individuals or groups who recruit and mobilize other community members to contribute and grow organically. However, since the first corporate-founded open source project (Mozilla) was launched in 1998, firms have begun to create or sponsor their own open source communities. Firms sponsoring open source communities typically do so as part of an intentional open innovation strategy (West and Gallagher, 2006). Because these communities are founded for strategic reasons, they are likely to differ from their autonomous counterparts. However, little is known about how corporate sponsorship affects how open source communities are designed and evolve. Too often the existence of technical communities is taken as a given, and the factors influencing their design unexplored (Hargrave and Van de Van, 2006).

If communities are an important vehicle for mediating firm interactions (Rosenkopf *et al.*, 2001) and potentially for innovation outcomes (von Hippel, 2005; Jeppesen and Frederiksen, 2006), then understanding such collaborations is crucial to any understanding of the roles that communities play in innovation. This research examines 12 sponsored open source communities, and contrasts them with prior research on autonomous communities. From this comparative analysis, we identified three design dimensions that corporate sponsors consider when designing open source communities: (1) intellectual property rights, (2) development approach and (3) model of community governance. We found that design decisions in these three areas created a specific participation architecture: that is, the socio-technical framework that extends participation opportunities to external parties and integrates their contributions. Much as architecture guides people in physical space, a participation architecture guides interactions and exchange in a community through the social, legal and technical capabilities offered to community members. While prior research has shown that a project's technical architecture can affect community participation (Baldwin and Clark, 2006; MacCormack *et al.*, 2006), there has been less appreciation for how community design choices can also affect participation.

By comparing the participation architectures that resulted from sponsors' design decisions, we identified two types of openness: transparency and accessibility. While transparency offered potential contributors the ability to follow and understand a community's production efforts, accessibility determined the degree to which external contributors could influence that production. In designing a community, sponsors were more likely to offer transparency than they were to offer accessibility to external community members. We found that sponsors faced a control vs. growth tension. To leverage the ability of communities to contribute to their firm's bottom line, sponsors sought to maintain control over the community's strategic direction. However, sponsors soon discovered that by restricting access to community processes, they limited their community's ability to attract new members and grow.

We contribute to the literature on open source communities, technical communities and firms and community collaboration in three ways. First, we identify some key distinctions between sponsored communities and autonomous communities that can help further research on firm–community collaboration and innovation. Second, we develop the construct of participation architecture and show how it is operationalized in a sample of open source communities. Third, we illustrate the "control–growth" tension that sponsors building communities face when making design decisions. Our research shows that participation in a community is determined not only by the technical architecture identified by Baldwin and Clark (2006), but also by the organizational structure that results from a sponsor's community building design decisions.

Firms and Technical Community Collaborations

To some extent, firms and technical communities have always collaborated (Allen, 1983) to create standards (e.g. Isaak, 2006), shared infrastructure (Bradner, 1999) and innovation outcomes (Hargrave and van de Ven, 2006) that are bigger than any one firm can achieve. Empirical work suggests three reasons why firms participate in technical communities.

First, there is increasing evidence that path-breaking innovations cannot occur without a community to interpret, support, extend and diffuse them (van de Ven and Garud, 1989; Anderson and Tushman, 1990; Tushman and Rosenkopf, 1992; Rosenkopf and Tushman, 1994; Christensen and Rosenbloom, 1995; Hunt and Aldrich, 1998; Hargadon and Douglas, 2001; Schoonhoven and Romanelli, 2001; Hargrave and van de Ven, 2006). Research on technologies such as medical devices, bicycles, computer hardware and electricity show that new technologies are shaped by human institutions that provide a context for the interpretation and use of such technologies (Bijker *et al.*, 1987).

Second, research on collective models of innovation (Allen, 1983; von Hippel and von Krogh, 2003; von Hippel, 2005; Hargrave and van de Ven, 2006) and on community technical organizations (Rosenkopf and Tushman, 1998; Rosenkopf *et al.*, 2001) shows that communities of lead users help firms not only in interpreting and applying new innovations, but in their creation and further development by refining new design iterations (von Hippel, 1988, 2005; Shah, 2006; Murray and O'Mahony, 2007). A large body of evidence suggests that firms in many industries (toys, entertainment, medical devices, manufacturing, sporting goods, music) benefit from contributions from community members (Franke and Shah, 2003; Lakhani and von Hippel, 2003; von Hippel, 2005; Jeppesen and Frederiksen, 2006).

Third, although it is not widely recognized, technical communities provide a vehicle to coordinate the work of both firms and individuals in developing new technologies and standards (Rosenkopf *et al.*, 2001; Mowery and Simcoe, 2005). Technical communities offer individuals leadership opportunities (Fleming and Waguespack, 2007; O'Mahony and Ferraro, 2007) and enhance their technical credibility (Hars and Ou, 2002; Lerner and Tirole, 2002). For example, participation in a specific public community (such as standardizing the http protocol as a member of the Internet Engineering Task Force) allows participants to both develop and advertise domain specific knowledge (Fleming and Waguespack, 2007).

The degree to which firms benefit from collaboration with technical communities can also depend on features of the technology itself such as the degree of modularity (Baldwin and Clark, 2000), or the number of complements, linking mechanisms or interface technologies required (Tushman and Rosenkopf, 1992). While practitioners have long argued that effective management of such communities can help support firm goals (Godwin, 1994; Armstrong and Hagell, 1996; Williams and Cothrel, 2000) little empirical work has been done in this area. To further our understanding of how firms and communities collaborate, scholars have recently turned their attention to specific types of technical communities, open source software development communities.

Early research on open source software communities focused on individual motivations to participate. Lerner and Tirole (2002) argued that contributors to open source communities participate in order to improve the visibility of their skills in the open labor market. However, subsequent empirical work suggests that volunteer contributors to community projects are equally likely to receive intrinsic benefits. Contributors to open source projects do so because they enjoy solving technical problems, they identify with the project's goals, they are interested in building their skills or simply want to improve the software for their own use (Hars and Ou, 2002; Hertel *et al.*, 2003; Lakhani and Wolf, 2003).

Since major corporations began incorporating community developed open source software in their products and services (e.g. MacCormack and Herman, 1999; West and Dedrick, 2001; Baldwin *et al.*, 2003), many communities have attracted donations of code and ongoing development participation from firms as well as individuals (O'Mahony, 2002, 2005; Dahlander and Magnusson, 2005). Because a community's output is publicly available, it limited the extent to which a firm's direct benefit from their investment in the community would remain unique to them. However, firms gained indirect economic benefits and competitive advantage by leveraging the widespread adoption of popular open source projects, leveraging the shared R&D investment of the community, and selling other goods and services necessary to provide a complete solution (West, 2003, 2007; West and Gallagher, 2006).

Some communities explicitly recognize firms as participants, while in other cases firms participate indirectly through employees (i.e. sponsored contributors) that represent the firm's interests (Dahlander and Wallin, 2006). Often, an open source community will create an informal or formal social structure to manage membership and joining (von Krogh *et al.*, 2003; O'Mahony and Ferraro, 2007), but little has been done to understand how these projects are governed (see Shah, 2006; Markus, 2007; O'Mahony, 2007 for recent exploratory exceptions).

In addition to collaborating with autonomous open source communities, a growing number of corporate and government sponsors have founded their own open source communities to meet either public or private objectives. West and O'Mahony (2005) distinguished between individually founded and organizationally founded open source communities, designating the former as "organic" and the latter as "synthetic". These distinctions emphasized the different character of these two types of community founders and their prospective growth strategies. While organic projects are founded by individuals and grow through grass roots communications, synthetic communities are founded by corporations and grow with more strategic direction. Since communities can evolve along different trajectories after their founding, in this paper, we refer to *autonomous* and *sponsored* communities to focus on their current governance structure (cf. Markus, 2007;

O'Mahony, 2007) as opposed to their founding state. For example, a synthetic community could begin as a sponsored community, but evolve to become fully autonomous—as was the case with Netscape's release of Mozilla and its subsequent transition to independence. By using the categories of autonomous and sponsored, our aim is to provide a more precise categorization to help scholars examine open source communities over their life cycle.

We define an *autonomous open source community* as one that is presently independent of any one firm and community managed (cf. O'Mahony, 2007). A community-managed governance system operates outside the reach of authority embedded in employment relations. Contributors to an open source project may be volunteers or may be paid by their employers to work on the project, but decision-making on the project takes place independently from the employment structure that guides the workplace. These projects may be supported by a non-profit foundation created specifically to support the project, but such foundations have little authority over their members (O'Mahony, 2005).

A *sponsored open source community* is one where one (or more) corporate entities control the community's short- or long-term activities. To refine these distinctions, in this study, we examine how sponsors approached the task of building an open source community and how these communities differed from their autonomous counterparts.

Methods

Given that there has been relatively little research on sponsored open source software communities, we examined why sponsors created open source communities and how their motivations affected the design of the communities they founded. Because sponsors understood that the design of an open source community affected a community's ability to attract contributors, and because this was widely regarded by sponsors as a sign of a successful open source community, we asked: how do community design decisions affect their ability to attract external participants? Since our informants frequently referred to autonomous communities as the inspiration for their own community building efforts, we analyzed the data in comparison with findings from prior research on a limited number of previously studied autonomous communities. For comparison purposes, we contrasted our findings with the earliest and most successful autonomous communities that had the greatest influence on the decisions of our firm sponsors: Apache and four major Linux-related communities.[1]

Research design. We adopted a grounded theory approach, which is well suited for phenomena that are emergent or poorly understood (Strauss and Corbin, 1990). Such an approach provides the broadest possible contextual information for understanding a phenomenon where there is not strong a priori theory (Edmondson and McManus, 2007). Thus, the study was guided by an inductive, qualitative approach using ethnographic methods.

---

[1] Because our focus was on sponsored communities, we did not gather primary data on autonomous communities, and thus our reference group was limited both in size and diversity. We believe that group was representative of early, successful autonomous projects but not of a broader population of autonomous projects that include both successes and failures.

Sample. To study sponsored open source communities, we employed a theoretical sampling approach to identify a wide range of possible structures and relationships (cf. Glaser and Strauss, 1967). We selected 12 sponsored open source communities founded between 1983 and 2004 that embraced elements of the open source model: an open source license, publicly available code and evidence of community on mailing lists. Since there was no single source available that could identify the population of sponsored open source projects, we drew upon our prior fieldwork in the open source field to identify such a population.

We excluded from our sample of study, those projects not perceived as open source by open source developers, such as Microsoft's "shared source" projects which provide a subset of open source intellectual property rights to a defined sub-group of customers (West and Dedrick, 2001) or "gated communities" that use open source development processes for a defined population without public release of intellectual property (Shah, 2006). Of the sponsored open source communities that we identified and met these criteria, six were founded by traditional large proprietary technology companies, five were founded by firms to commercialize the open source software produced by the community and one was sponsored by a non-profit organization set up specifically for that purpose. Table 1 presents descriptive information on the 12 projects selected.

Data. As this was exploratory research, we interviewed one to six informants from each sponsoring organization. After selecting communities, informants were identified from project web pages and industry conferences. Semi-structured interviews were 60–90 minutes long, and focused on understanding how sponsors approached the prospect of building community. More specifically, interviews covered the following domains: (1) when

Table 1. Sample of sponsor-founded open source communities

Founding date	Project	Type of software	Sponsor	Sponsor type	Interviews
1983	Sendmail	Electronic mail server	Sendmail	Open source startup[a]	1
1990	Berkeley DB	Database	Sleepycat	Open source startup[a]	2
1995	MySql	Relational database	MySQL AB	Open source startup	1
1997	PHP	Web scripting language	Zend	Open source startup	2
1998	Mozilla	Web browser	Netscape	Proprietary IT firm	3
1998	Jikes	Java compiler	IBM	Proprietary IT firm	2
1999	Darwin	Operating system kernel	Apple	Proprietary IT firm	3
2000	OpenOffice	Office productivity suite	Sun Microsystems	Proprietary IT firm	4
2001	Eclipse	IDE/application platform	IBM	Proprietary IT firm	6
2002	Helix	Media streaming	RealNetworks	Proprietary IT firm	2
2003	Chandler	Information manager	Open Source Applications Foundation	Non-profit corporation	3
2004	Sugar	Customer management	SugarCRM	Open source startup	1
				Number of interviews	29

[a]Originally university sponsored.

and how sponsor founded an open source community; (2) how they prepared for and designed an open source community; and (3) their experience thus far in building community. Most of the interviews ($N = 23$) were conducted face to face, with the remainder ($N = 6$) over the phone. These visits were followed by an analysis of the community's website and follow-up questions posed to the informants as appropriate. We collected our data from early 2002 through mid-2005.

Data analysis. We analyzed the interview transcripts, coding relevant observations and compared and contrasted our interview notes. Emergent themes in the data were synthesized in research memos written in an ongoing research log. As our research progressed, we contrasted our interim findings with prior research on autonomous communities to better understand the distinctions between the two.

Findings

While prior research on open innovation has found that open approaches to developing communities can vary in the degree to which they are "open", our informants taught us that when forming a community, there were two distinct types of openness: *transparency* and *accessibility*. Transparency allows outsiders (i.e. non-sponsors) to understand what is happening and why—and, in the case of an open source community, allows use of the community's final product, the source code. Accessibility allows external participants to directly influence the direction of the community to meet their specific wants and needs, regardless of whether the external party is a hobbyist or an organizational adopter. In some cases, external contributors could be sellers of goods and services that might either compete with or complement the sponsor's business.

How were these two different types of openness provided? By comparing our data across 12 sponsored open source projects, we found that the design choices made by sponsors of open source communities could be categorized into three dimensions: (1) the organization of production; (2) community governance; and (3) intellectual property. We identified these conceptual categories by examining how sponsors' design choices affected what community participation. Decisions with regard to the organization of production shaped how code development would take place, while community governance decisions shaped the processes by which decisions were made within the community; and intellectual property design decisions affected the allocation of rights to use the community's output. Table 2 shows how these three design dimensions compare with those of proprietary software development, the approach historically used by companies in the software industry.

After noting the importance that our informants placed on designing community forums for participation, we realized that, when considered together, these three dimensions of community design formed the basis of a participation architecture. We define a participation architecture as the socio-technical framework that extends opportunities to external participants and integrates their contributions.[2] A participation architecture guides

[2] The concept of an "architecture of participation" was initially articulated by O'Reilly (2005) as a set of heuristics that encourage participation and innovation. The "participation architecture" construct we develop here is intended to meld that earlier usage with an interorganizational analog to the "technical architecture" of Baldwin and Clark (2006).

Table 2. Mapping forms of openness against dimensions of open source

		Form of openness		
		Transparency	Accessibility	*Proprietary model*
	Production – the way that the community conducts production processes	Ability to read code and observe or follow production processes	Ability to change code directly	Production remains within a single corporation
Dimension of participation architecture	**Governance** – the processes by which decisions are made within the community	Publicly visible governance, observers can understand how decisions are made	Ability to participate in governance	The corporation makes all decisions at its own discretion
	Intellectual property – the allocation of rights to use the community's output	Rights to use code and access source code	Ability to reuse and recombine code in the creation of derivative code	Limited use rights are granted by the corporation for a licensing fee

interactions and exchange in an online community and encompasses the social, legal and technical capabilities offered to community members. While sponsored communities were by some measures less open than most autonomous communities, in all cases, the decision to create an open source community was inherently a more open approach than a proprietary software development approach.

Within these three design dimensions that we categorized, we identified 11 design parameters chosen by the founders and subsequent leaders of the communities in our sample (Table 3). Both transparency and accessibility were relevant across each of the three dimensions of an open source project. After coding our data to understand whether a design parameter affected the organization of production, governance or intellectual property, we revisited it to determine whether that design choice enabled transparency or access.

We found that the degree to which sponsors offered transparency or accessibility in these key areas affected sponsors' ability to attract external participants and grow these communities. By examining sponsors' community design decisions, we discovered that one of the primary challenges sponsors faced was how to manage the tension between controlling the community in order to leverage their investment in it and opening up access to the community in order to attract greater growth of participants.

Managing the Tension between Control and Openness

One key difference between autonomous (community managed) and sponsored open source software projects is that the sponsors of open source projects faced a fundamental tension between two conflicting goals. On the one hand, sponsoring an open source project was intended to advance the goals of the sponsoring organization. Sponsoring an open source project required significant investment in preparing the code, hosting the site, providing introductory materials and marketing the new opportunity. As such, community

Table 3. Design parameters for autonomous and sponsored open source communities

	Production			Governance				Intellectual property			
Design parameter	Live code access	Public commit process	Sub-project creation	Non-profit foundation	Member-ship	Member fee	Community release authority	Code owned by foundation	Subproject ownership	Software license	License type
Autonomous											
Apache	X	X	X	X	Individual	No	X	X	Foundation	Apache	Permissive
Gnome	X	X	X	X	Individual	No	X	X	Foundation	GPL	Reciprocal
Debian	X	X	–	X	Individual	No	X	X	–	GPL	Reciprocal
Linux	X	[a]	X	[a]	–	–	[a]	[a]	Community	GPL	Reciprocal
Linux Standard Base	X	X	X	X	Firm and individual	X	X	X	Foundation	GPL	Reciprocal
Sponsored											
Eclipse*	X	X	X	X	Firm and individual	X	X	X	Foundation	Eclipse	Partly reciprocal
Mozilla*	X	X	X	X	Planned	–	–	X	Foundation	Mozilla, GPL	Partly reciprocal
OpenOffice	X	X	X	Planned	Planned	–	–	–	Sponsor	LGPL	Partly reciprocal

*Evolved from sponsored to autonomous.

Table 3. Continued

Design parameter	Production			Governance				Intellectual property			
	Live code access	Public commit process	Sub-project creation	Non-profit foundation	Member-ship	Member fee	Community release authority	Code owned by foundation	Subproject ownership	Software license	License type
Helix	X	X	X	–	–	–	–	–	Sponsor	RPSL, GPL	Reciprocal
Sugar	X	–	X	–	–	–	–	–	Participants	Sugar[a]	Partly reciprocal
PHP	X	X	–	–	–	–	–	–	–	PHP	Permissive
Chandler	X	–	–	a	–	–	–	–	–	GPL	Dual
Jikes	X	–	–	–	–	–	X	–	–	CPL	Partly reciprocal
Darwin	X	a	–	–	–	–	–	–	–	APSL	Permissive
MySql	X	–	–	–	–	–	–	–	–	GPL	Dual
Sendmail	–	–	–	–	–	–	–	–	–	Sendmail[a]	Dual
Berkeley DB	–	–	–	–	–	–	–	–	–	Sleepycat	Dual

Communities are listed in order of decreasing level of overall openness. Notes on specific communities:

- Linux: allows external committers but the process is not public; has an affiliated non-profit that influences but does not control the project; release authority is vested in key individuals; code in the Linux kernel is licensed by contributors to the community but none is owned by the community.
- Chandler was founded by a non-profit.
- Darwin: outside commit rights allowed on parallel, experimental code repository.
- Sendmail, Sugar: license is not approved by the Open Source Initiative.

sponsors sought to maintain some degree of control over the project to assure ongoing alignment between their investment in the community and related product goals.

On the other hand, the provision of source code under an open source license was an inherently open approach intended to win greater external participation and technological adoption. In some cases, sponsors sought adoption from prospective users (cf. West, 2003); in other cases, they sought adoption from producers of complementary products or even direct competitors.[3]

We found that for the most open communities, the participation of external parties provided sponsors with both direct benefits (such as code contributions and bug reports from participants) and indirect benefits (such as marketing and adoption benefits from their open approach). For the most closed communities, sponsors thought that the primary benefit they received from creating an open source community was not from direct community contributions, but instead from increased public awareness, accelerated low cost distribution and reduced costs of marketing.

The CEO of one open source startup explained how expensive commercial marketing channels were relative to the marketing benefits that could be derived from an open source community.

> Every dollar you give [proprietary competitor], 70 cents to goes to fund the sales and marketing efforts, and maybe 11–14% actually goes to pay the engineering salaries that write the code.... The barriers to entry into this marketspace isn't building a better product, it is having $50–60 million a year just to blow on sales and marketing. And that's really a shame. I just thought that was really frustrating from someone that is an innovator, someone that wants to compete by building a better product [rather than] on just the sheer economics of sales and marketing.

While sponsors shared a surprising degree of agreement on their motivation to create an open source community and the most immediate resulting benefits, they made very different design choices with regards to the type and degree of openness with which they were most comfortable. Our informants taught us that there were two kinds of openness to consider when designing an open source community: transparency and accessibility. Both transparency and accessibility were relevant across each of the three dimensions of an open source project (Table 2).

Transparency. Transparency meant that the code was publicly available, that most of the software production process was discussed on public mailing lists or discussion boards, and that the software release cycle and goals were also provided on the community website. While this was fairly unproblematic for static information about a project, the transition to a transparent production *process* took some getting used to for most sponsors. After receiving complaints that they were not being transparent enough in their development process, one community sponsor reevaluated what needed to be public and what needed to

[3] Comparing the size of the community (or amount of community participation) between our 12 communities is problematic because the product category for some projects (e.g. Mozilla) has much bigger potential audiences than others (e.g. Sendmail). Such comparisons are made more difficult by the relative scarcity of competing open source projects within the same category. Thus we rely on each project's sense of its growth relative to the inherent potential for such growth, including the statements by some informants that they made choices to restrict openness that they knew limited the growth of contributions or other forms of participation.

be private, deciding to err on the side of transparency. In making that decision, they decided to limit only private discussions only to those involving third parties—but this required changing ingrained habits. As its community manager explained:

> It [becoming open] took a while. We talked about it before. Then we got a complaint from someone working with us, one of the contributors saying I think I'm missing stuff. I don't know exactly how justified that particular complaint was. I am not sure that I agreed with that.... Then we looked at the development threads, and said, "Well what of this really needs to be private?" "The third party stuff", and then we said, "Okay let's do it [be open]." So some of it I think was timing, as people get more used to being open. We went through a very similar thing at [another sponsored project], killing the internal lists.... So we renamed them with long, ugly, awkward names. Of course you might get auto-fill or something, but if you just naturally type, you got the public list. If you wanted to do something private it was hard instead of easy. But because a lot of that wasn't malfeasance, it was just the habit of doing what you do.

This sponsor renamed internal mailing lists to encourage developers to default to public lists and change the base of operations from private to public: a technological change to instigate a cultural and process change. While the sponsor reported that there were one or two human resources issues that ended up on the list that probably should not have, the changes were largely viewed as positive for both the organization and the community, and resulted in the development of a new reporting tool that could simultaneously be used for both public and private purposes.

Accessibility. The second type of openness sponsors considered when designing a sponsored community was the degree of accessibility—the amount of control sponsors would relinquish to the community. An accessible community not only provides visibility, but allows some outsiders to gain access to either the project's code repository, community planning processes or strategic decision-making. Much like a community founded by individuals, the decision to provide access at different levels was usually phased and based on community members' demonstrated competence and evidence of contribution. However, sponsors supporting even the most accessible communities were likely to retain control over rights allocation rather than devolving it to the community. Two sponsors delegated control of code commit rights for subprojects to external parties, but retained control of higher level decision-making.

In this way the organization of production and governance became linked (to be discussed further in the next section). In doing so, sponsors explicitly recognized that granting accessibility triggered the loss of some corporate control in a way that a purely transparent model did not—particularly when external participants were from other (possibly competing) companies. In a community that did not yet have a formally approved set procedures for decision-making right allocation, these issues were constantly negotiated among contributing firms. As another community manager explained:

> If you are building an Open Source project, which has several commercial players at a significant level, you do a lot of negotiating and a lot of figuring out, "Well what is the right milestone schedule? ... What is the direction of the code base? And what time frame? Who can contribute

> code? How do you decide what code is good enough? Who controls, or how do you control the relationship of the [project] release to those of commercial products?"

In general, sponsors who most valued outside code contributions were more likely to offer accessibility to outside community members, recognizing that only by devolving some level of control, could they hope to attract the most talented programmers outside the firm.

All sponsors worked to achieve significant transparency in their open source communities, but sponsors varied considerably in the importance they placed on providing accessibility to external parties. This distinction provides a more nuanced understanding of the tension between openness and control. To make this abstract tradeoff more concrete, we identified three specific community design parameters that affected the degree of transparency and accessibility that sponsors provided.

Organization of Production

When forming an open source community, sponsors borrowed heavily from the tools and techniques pioneered by autonomous open source communities. This included general-purpose online tools such as web pages and particularly email discussion lists. However, most also used some of the tools specifically developed and refined for open source software production, notably the CVS source code control system and the Bugzilla error tracking database (Robbins, 2005).

On the technical side, one key enabling element is modularity. Baldwin and Clark (2005) show how a modular architecture offering design options increases a developer's incentives to join an open source community and remain involved. Architectures that are modular allow developers to focus their talents on specific modules without having to learn the whole system (Baldwin and Clark, 2005). By maintaining compatibility with design rules within modules developers can self-select the modules they know best, reducing participant learning curves and thus lowering the cost to participate (Baldwin and Clark, 2000).

Sponsors were aware of how the degree of modularity could affect potential barriers and costs to participation. Thus, sponsors invested significant resources in creating or increasing modularity, consistent with what MacCormack *et al.* would predict (2006). For example, the founders of one project rewrote its version 3.0 to create more modular interfaces than previous versions—to make code development by others easier and facilitate modular extension by community members. Another dual licensed project's highly interdependent architecture provided high performance but made it nearly impossible for outside participants to become proficient with the code—thus inhibiting potential community contributions. In contrast, the sponsor of one of the newer communities emphasized that "we wrote the code knowing it would be read"; they provided code reviews and other quality improvements to explicitly attract outside contributions within months of its first release. Another sponsor bragged that his was the "best documented open source project".

Overall, the degree of modularity, associated dependencies and the quality of code documentation affected the ability of outside members to understand the code well enough to contribute. However, many of the measures available for individual projects were not comparable. In addition to the technical architecture of the code, the organization of production includes control of the processes by which individuals participate in the community's production process. These social measures are not necessarily correlated to a project's technical design: for example, highly modular code can still be tightly controlled by a single firm. Thus, a project's technical architecture is one subset of a community's participation architecture. As the rest of the paper will show, to more fully understand how both social and technical attributes affect the opportunities for others to participate requires consideration of all three design parameters—the organization of production, governance and intellectual property. We identified three design parameters that provided contributors with transparency and accessibility to production processes:

1. *Live code access* provides transparency by offering the community the chance to review the most recent "live" version of source code on the community website—which is more likely to have bugs than a finished product. Nearly all of the sponsored communities (10/12) allowed external participants to anonymously access the most current source code, subject to the sponsor's license terms. External contributors were thus able to follow the community's development cycle and contribute bug reports if they were so inclined.
2. *Public commit process* refers to the opportunity for community members to become directly involved in the production process by earning (through demonstrated technical proficiency) the right to directly commit software changes to the community repository. While all of the autonomous communities provide such accessibility, only some (5/12) of the sponsored communities did so. Seven of the sponsored communities did not publicly explain how one could go about acquiring commit rights to their projects, but most of them (6/7) encouraged people to send code patches via email.

   The lack of information about gaining committer rights is not necessarily an inhibitor for participation but it limits the status and influence that a contributor can achieve. As one community member explained, "people can be phenomenally valuable contributors without having access for a long time. Somebody else can check their code, but people don't like it. It is seen as a mark of belonging." Sponsors that did not provide commit rights recognized such rights would increase participation but were unwilling to relinquish that much control over their code.
3. *Subproject creation* is a mechanism by which a community based on the sponsor's original code can grow to assume new functionality or new directions. We defined subprojects as new "start-up" projects that were allowed to govern themselves and address unmet needs independent of the larger community; allowing creation of such subprojects provides accessibility by decentralizing control over growth and innovation in the community. While all of the autonomous projects allowed community members to autonomously propose subprojects based on their own initiative, only 5 of the 17 sponsored communities did so.

Assessing openness. Source code access offers potential external community participants transparency into the development process which can help them learn how the code is developed. Without awareness of the community's production process, the learning curve to make a meaningful contribution and thus gain membership or access within the community will be limited. The ability to create subprojects offers access to external parties by providing them opportunities to shape the future direction of the project. This type of access sends a powerful signal as to the ease with which external contributors can get new ideas proposed and accepted. Allowing individuals the right to earn commit rights offers the ultimate degree of accessibility—the ability to make direct contributions to the code. As Table 3 shows, more sponsors were more likely to provide transparency in their production processes than they were to provide accessibility.

Governance

We define openness in open source governance as the amount of decision-making control that sponsors relinquished to the community. For many in our sample, divesting some degree of control was as much a legitimating strategy as it was a recruitment strategy. To attract talented contributors, sponsors thought they both needed to acquire legitimacy in the open source community at large, and provide skilled participants the opportunity to take on greater responsibility in leadership roles. As one community founder from a Fortune 100 firm explained:

> [P]art of the message behind open source is that it is open and the community makes the decisions. And if what we actually did was said that the community makes decisions, but in practice, [the sponsor] makes all of the decisions then they would say "well this is not real".

To provide both leadership and legitimacy, some communities have a formal concept of membership, vesting members with key governance decisions, and are thus more accessible to participants. For other communities, de facto control remains with founding individuals (due to superior legitimacy or technical knowledge) or with founding firms (that provide the bulk of ongoing resources). The design levers available for managing openness in governance include:

1. *Non-profit foundations.* All five individually founded communities and two sponsor-founded communities created a formal, legal, non-profit foundation to help manage community governance and assets; however, for historical reasons, the Linux foundation plays a more limited role than the others. As O'Mahony (2003, 2005) identifies, such foundations provide institutional permanence independent of any one individual, as well as legal status to negotiate with external entities (largely for the provision of resources). Because the creation of a non-profit foundation requires a board of directors and regular meetings, the introduction of such organizations also increased the transparency of control of the community assets.

2. *Membership.* Four out of five of the individually founded open source communities have formal processes by which an individual is recognized as a member; one community recognized both individuals and firms as separate classes of members. Only one sponsor-founded community project created a non-profit foundation with a membership base, while two others were under development at the time of our study. After offering external contributors membership rights, this sponsor experienced unprecedented growth in participation. Communities with a membership base provide members with some voice in formal governance matters (typically through annual elections and the right to vote on project-wide decisions such as license or name changes), while non-membership communities remain either fully sponsor-controlled or retain ad hoc governance mechanisms.

3. Membership fee. Two communities (one individually founded and one sponsor founded) obtain funds from interested firms by selling memberships at a range of prices. In general, membership fees were not adopted by sponsors trying to create communities, but by those sponsors trying to create commercial ecosystems.

4. Release authority. The ultimate test of an online production community is who makes the final production decisions. For an open source community, this decision occurs when software is released. This authority may vest in the members, the affiliated foundation or remain with the sponsor or individual founder. Again there is a dramatic difference in openness between autonomous and sponsored projects: the community holds authority in all but one autonomous community (Linux), but only in two of the 12 sponsored communities. Such limited accessibility appears to reflect the sponsor's desire to align the features, quality and schedule of open source releases to its commercial goals.

Assessing openness. By creating a membership organization and the opportunity to elect leaders, a few sponsors offered potential contributors the ability to develop a sense of belonging and become more vested in the community's future. The ability to gain "membership status" was viewed as a motivational and recruitment tool. Sponsors who adopted these community design features did so with the belief that divesting some control was necessary in order to attract talented contributors. However, most sponsors did not create an independent form of governance, retained exclusive release authority and final say on all key community decisions.

Intellectual Property

Sponsored open source communities faced a potential barrier to external participation not found in their autonomous counterparts: the fear that the founder's desire to profit from community-based production would limit the benefits that accrued to contributing members. West (2003) refers to this as a producer's inherent tradeoff between winning adoption of a technology and appropriating the returns from that technology. For open source communities, the key attributes of this tradeoff were associated with the ownership and licensing of the community-produced software:

1. Content ownership. Most communities that created an affiliated foundation also vest ownership in the foundation (Linux an exception), as did two sponsored communities. For the other sponsored communities, ownership remains with the sponsor. Ownership of the content by a foundation provides a credible assurance to community participants that the code will remain available to participants in perpetuity (O'Mahony, 2005). The direct ownership of the code by sponsors in most of the communities was a clear direct and symbolic measure of their design to maintain ongoing control over the terms by which outside participants access the code.

2. Subproject ownership. For the projects that allowed creation of new subprojects, the conditions of access were not necessarily the same as for the core code. For the five sponsored projects that allowed subproject creation, in four cases, ownership of the subproject output was the same as for the core project (two owned by the sponsor, two owned by the foundation); for one sponsored project, the output was owned by the contributing community members.

3. Software license. Ten of the 12 sponsored communities use one of the more than 50 licenses approved by the Open Source Initiative (OSI). As Lerner and Tirole (2005) and others have noted, the most popular open source license is the GNU General Public License (GPL). This widespread use and popularity among potential participants influenced the decisions of two sponsors—Mozilla and Helix—to offer their product under both their own license and the GPL, while other sponsors mentioned the importance of using a license deemed compatible with the GPL or the Lesser GPL (LGPL). Two sponsors used licenses that are as yet unapproved by the OSI. One community (Sendmail) was founded using an approved license (BSD) but the sponsor subsequently chose to release a major update under a non-approved license that discriminates against for-profit users. The last (Sugar) created its own license (adapted from the Mozilla Public License) that required publicity for their technology when used by service providers.[4]

4. License type. Open source licenses such as the BSD or Apache license allow recipients to use the code largely without restriction, while "free software" licenses (notably the GPL) compel recipients to return any modifications or changes (West, 2003). Rosen (2004) classifies these two types as "permissive" and "reciprocal"; a few licenses (such as the LGPL) impose reciprocal obligations on part but not all of the code. Four of the sponsored communities in our sample use a dual license strategy to implement price discrimination, with non-profit users selecting the free reciprocal license (i.e. the GPL) and for-profit users customarily paying to use the software without the reciprocal obligations (cf. Välimäki, 2003).

Assessing openness. The ownership of code was the most dramatic difference between autonomous and sponsored projects, in that the sponsor in nearly all cases retained ownership of the core (if not subproject) code. A subset of the sponsors used the objectionable restrictions of the reciprocal license to provide revenues through a dual license policy, but otherwise the sponsored and autonomous communities followed similar license policies, either using standard or their own open source licenses.

Discussion

We defined and contrasted two different types of open source software communities: those sponsored by corporate organizations and the more traditionally studied autonomous (community managed) communities. Our study had two research questions: how did sponsors design open source software communities in the hopes of attracting external participation, and how did this differ from the design of autonomous-based communities?

By studying the design decisions that sponsors made when creating a community, we identified three dimensions that affected participation: (1) the organization of production, (2) governance and (3) intellectual property. In doing so, we showed that the participation architecture of a technical community is determined not only by its technical architecture,

[4] Sponsors of the two unapproved open source licenses (Sugar, Sendmail) argued that their licenses largely met the requirements of the Open Source Initiative (OSI), although they did not submit them for OSI approval. Their use of unapproved licenses they termed "open source" stimulated occasional controversy within the open source social movement, but did not seem to impair the effectiveness of their community strategies. In December 2007, SugarCRM solved this problem by releasing its free software under GPL Version 3, an OSI-approved license.

but also by community design decisions made by the community's leaders. While modularity in the technical architecture remains important to enabling participation by reducing the learning curve or cost of entry (e.g. Baldwin and Clark, 2006), the aspects of community design that we identified are also critical to attracting and enabling participants primarily because they shape the landscape of opportunities extended.

We showed that sponsors' community design decisions on these three dimensions reflected the inherent tension between two conflicting goals. On the one hand, firms wished to retain control over technologies fundamental to their business success. On the other hand, providing the opportunity structure for others to participate was a prerequisite for gaining the benefits from developing an external community. Thus, when designing a participation architecture, firms mediate between surrendering control and offering opportunities for outside participation that could lead to community contributions and growth.

The Role of Participation Architecture in Growing Sponsored Communities

We discovered that before designing their own open source software communities, our informants studied well-known autonomous communities in some detail and made frequent reference to them. Thus, in the presentation of our findings, we compared the findings from our sample to some of these well-known autonomous communities.

Based both on qualitative data from our informants and online data from the communities in our sample, we found a fundamental tension unique to sponsored communities: while sponsors recognized that the key to attracting and retaining participants to their communities was to provide unfettered opportunities for contribution, they had an interest in retaining some controlling influence over the communities they founded to ensure these communities remained aligned with corporate strategy. Managing this tension was a pervasive concern and illuminates some of the challenge in using external communities to pursue open innovation. After identifying the tension between openness and control, we identified more precisely how sponsors reconciled this tension with 11 specific community building design parameters that cluster across three dimensions.

From our data, we found that the sponsor-founded communities could be classified into one of three distinct groups as sorted in Table 3. The first group of firm-created communities either had achieved or were seeking levels of community participation comparable to those of individually founded communities; in fact, two of the communities (Eclipse, Mozilla) transitioned from corporate sponsored to become independent autonomous communities.

At the other extreme, a second group of communities (the three firm-sponsored dual-licensed communities) offered what we term a "fishbowl" development pattern—with the sponsor offering transparency to outsiders, but not accessibility to software development. A third group of communities lay somewhere in between: experimenting with the provision of access but not willing to give up key points of control.

Strikingly, sponsors were far more likely to provide transparency than they were accessibility, despite the possibility that a more controlled governance structure offered fewer opportunities for leadership and could thus reduce the sponsor's ability to recruit contributors. Community design decisions to provide either transparency or accessibility had very different effects. Transparency was cited by informants as critical to aiding adoption of the software: a key goal of all sponsors. The effects of accessibility were more mixed: while accessibility could potentially enhance the volume and quality of contributors to a project, it could also compromise sponsors' control over production. For when development was made fully accessible to external parties, more parties to decision-making created new dependencies and coordination costs for software that was critical to firm product lines.

However, there was no direct evidence of the direction of causality between the provision of accessibility and external community participation. Firms offering less accessibility could be motivated by a need to retain control, a belief that there was no benefit to doing so (because the community would not help in production anyway), or perhaps diminished expectations of external participation that became a self-fulfilling prophecy. Accessibility is only one of the factors that drive participation, as the high rate of participation (and adoption) for the tightly controlled MySQL community would suggest.

Most of the sponsored communities produced software of interest to a large potential audience of user-adopters. Except for the dual-license software communities, there seemed to be little relationship between a sponsor's license choices and the overall accessibility of the community. However, communities that were less accessible (Darwin, MySQL, Sendmail, Berkeley DB) seemed to be due to a sponsor's stronger need for control due to a greater fear of cannibalization of core revenues. Conversely, four of the most accessible sponsored communities (Eclipse, Mozilla, OpenOffice, Helix) produce software facing intense competition against a well-funded proprietary alternative, and thus these sponsors were most concerned with attracting external collaborators to aid in production and adoption. The participation architecture offered by these communities most closely resembled autonomous communities.

While a few sponsored communities sought outside participation by emulating key accessibility characteristics of autonomous projects (Apache was cited as a notable model), informants suggested that both creating accessibility and attracting significant external resources was a long and difficult process. Sponsors often approached the challenge in phases, offering transparency while preparing for accessibility. As one community manager trying to guide his community from a transparent model to a more accessible one explained, "the transparency and the communication of what is happening is a prerequisite for almost anything else".

Contrasting Sponsored and Autonomous Communities

Autonomous open source software communities have received a great deal of empirical and scholarly attention within the last decade. However, there has been very little research on corporate-sponsored open source communities, on how they differ from autonomous ones, or on how such communities contribute to a firm's open innovation strategy. This research takes a first step towards answering these questions. By comparing sponsored communities and autonomous communities we found some important commonalities. Both offer access to code that is guaranteed by an open source license, which fits the definition set by the OSI. Both also offer a high degree of transparency of access to that code—without which the rights to use the code would be useless.

However, our study showed that sponsored open source software communities are fundamentally different from autonomous communities in the potential for goal conflict between sponsor and community members. Although both sponsors and members seek widespread adoption, the primary goal of a corporate sponsor is profiting from its investment, while the goal of an open source community would be improving the capabilities of the shared technology.

To gain interest from a community of contributors, sponsors needed to at least provide transparency. The openness of sponsored communities differed most in terms of accessibility, with most sponsors retaining privileged (monolithic) rights for some portion of the community's decisions. In a few open cases, the sponsor shared some control with the community—and when sponsors relinquished more control to the community, those sponsored communities were transformed into autonomous ones.

As a consequence, we also found a dramatic difference between most sponsored and autonomous communities in terms of design decision related to accessibility, particularly in terms of governance. Governance of autonomous projects was largely pluralistic, shared widely among community members, whereas the ultimate decisions of sponsored communities were (with rare exceptions) controlled by the sponsor.[5]

The dichotomy is not complete, because not all autonomous open source projects provide full accessibility. Raymond's stylized typology (1999) of "The Cathedral and the Bazaar" contrasts the tightly controlled BSD projects with the more open Linux. However, today the "cathedral" archetype is relatively rare: Raymond's criticism (and the success of Linux) have meant that successful autonomous projects have largely followed Linux in granting accessibility to potential contributors. However, as a practical matter the importance of community contributions constrains the accessibility decisions of autonomous communities more than sponsored ones: independent communities that don't attract contributions will have trouble producing new software, while sponsors can (and do) sustain communities with their own resources—as happened with MySQL and Berkeley DB in our sample.

Future Research

Our comparative analysis of sponsored open source software communities and identification of the theoretical constructs that affect community design should enable future comparative work in community innovation. Our inductively generated framework can help scholars explicate and articulate differences and similarities across technical communities involved in the wide range of production of shared information goods, whether such goods are software, reference data (e.g. Wikipedia) or a travel guide (World66). Future research would do well to quantify how the design of a participation architecture affects a peer production community's growth trajectory.

Our study focused on corporate-sponsored open source communities. However, we recognize that an increasing number of organizations such as private non-profit foundations, governments and even transnational organizations sponsor technical communities. We would expect that communities sponsored by government or non-profit

---

[5] We use as our "control" group those large, successful pluralistic autonomous projects best known to our informants. Like our informants, we are thus drawing inferences from best practices rather than a cross section of autonomous projects, which limits our ability to draw contrasts. For example, those autonomous projects that are less successful in attracting an external community would have less developed formal governance than the successful projects listed in our control group.

actors would be more likely to favor public good ahead of the sponsor's pecuniary gain, but face similar tensions between maintaining control and attracting community participation and growth.

Firms have long sponsored external communities of users (such as chat rooms or bulletin boards) to provide communication with users, to both diffuse new technology and obtain user feedback. Sponsorship of external communities has been used as a source of open innovation, whether in musical instruments (Jeppesen and Frederiksen (2006), computer games (Prügl and Schreier, 2006; West and Gallagher, 2006) or sporting goods (Franke and Shah, 2003). Open source software has been held up as an exemplar of the process of user-contributed or open innovation (von Hippel, 2001, 2005; West and Gallagher, 2006). Certainly, the proliferation of autonomous open source projects has brought a raft of experimentation and proliferation of community forms. However, the growing popularity of sponsored communities suggests that firms can also sponsor open source communities and attract external participants. The question that remains open is under what conditions are firms more likely to sponsor such communities? Are these communities alternatives or complementary to in-house software development activities? How do they relate to a sponsor's broader research and development and outreach strategies?

While we believe our study makes an important contribution to our understanding of sponsored open source communities—and open source collaboration more broadly—important questions remain unanswered. For example, how does the creation of such communities affect firm business models and practices more generally? There are many opportunities for further research on sponsored communities. One area is the long-term impact of the tension between sponsor and participant goals—are external participants eventually discouraged by the sponsor's ongoing control, or do sponsors increase accessibility over time as they learn how to do so without surrendering full project control.

There are also questions about the changes of sponsorship over time. The imprinting of an organization at founding is certainly important (cf. Stinchcombe, 1965), but control over the community can change over time. Communities may transition from autonomous to sponsored, as has happened when a community founder (usually a hobbyist-programmer) forms a company to monetize the value of the code. Or they may transition from sponsored to autonomous, often as part of a larger transition from a proprietary software project to a sponsored open source community to an autonomous community. Both Mozilla (now Firefox) and Eclipse went through such transitions, but they have received little empirical examination. What affects the evolution of such projects? What consequences do these changes have for the code, the community and for innovation in general?

Our work is suggestive but hardly definitive on the role of technical aspects of openness. Consistent with Baldwin and Clark (2005), our informants suggested that technical structure (modularity) was one key aspect, while the other (along the lines of software engineering practice) was coding style. But are these categories mutually exhaustive? What about design elements (such as well-documented programming interfaces) that span both categories? How would these be measured? The next step would be to test the predictive value the constructs we identified. Would our social/structural measures predict participation? Or would technical openness have greater predictive value? Or is it some other factor, such as product quality, the size of the target market or price of the existing alternative?

Finally, as with any study conducted in a single industry context, there are opportunities to verify the generalizability of the findings—in this case, whether the sponsorship processes identified in open source software apply to other types of distributed content generating or innovation-focused online communities. There are a host of peer production and content generation communities that have flourished in recent years—some non-profit and some commercial, however how a community's participation architecture affects either community or commercial growth has not been teased apart. Our hope is that this framing offers a starting point.

Acknowledgements

Earlier versions of this paper were presented at the Academy of Management, Hawai`i International Conference on System Sciences, and European Academy of Management conferences. The authors thank the guest editors, two anonymous reviewers and especially the EURAM participants for their helpful suggestions.

References

Allen, R. C. (1983) Collective invention, *Journal of Economic Behavior and Organization*, 4(1), pp. 1–4.

Anderson, P. and Tushman, M. (1990) Technological discontinuities and dominant designs: a cyclical model of technological change, *Administrative Science Quarterly*, 35(4), pp. 604–633.

Armstrong, A. and Hagell, J. (1996) The real value of online communities, *Harvard Business Review* (May), pp. 134–141.

Baldwin, C. Y. and Clark, K. B. (2000) *Design Rules, Vol. 1: The Power of Modularity* (Cambridge, MA: MIT Press).

Baldwin, C. Y. and Clark, K. B. (2005) The architecture of cooperation: does code architecture mitigate free riding in the open source development model? Harvard Business School Working Paper Series No. 03-209, June.

Baldwin, C. Y. and Clark, K. B. (2006) The architecture of participation: does code architecture mitigate free riding in the open source development model?, *Management Science*, 52(7), pp. 1116–1127.

Baldwin, C. Y., O'Mahony, S. and Quinn, J. (2003) IBM and Linux, Harvard Business School Case 9-903-083.

Bijker, W. E., Hughes, T. P. and Pinch, T. (Eds) (1987) *The Social Construction of Technological Systems* (Cambridge, MA: MIT Press).

Bradner, S. (1999) The Internet Engineering Task Force, in: C. DiBona, S. Ockman & M. Stone (Eds), *Open Sources: Voices from the Open Source Revolution*, pp. 47–52 (Sebastopol, CA: O'Reilly).

Christensen, C. M. and Rosenbloom, R. S. (1995) Explaining the attacker's advantage: technological paradigms, organizational dynamics and the value network, *Research Policy*, 24(2), pp. 233–257.

Dahlander, L. and Magnusson, M. G. (2005) Relationships between open source software companies and communities: observations from Nordic firms, *Research Policy*, 34(4), pp. 481–493.

Dahlander, L. and Wallin, M. W. (2006) A man on the inside: unlocking communities as complementary assets, *Research Policy*, 35(7), pp. 1243–1259.

Edmondson, A. and McManus, E. (2007) Methodological fit in management field research, *Academy of Management Review*, 32(4), pp. 1155–1179.

Fleming, L. and Waguespack, D. M. (2007) Brokerage, boundary spanning, and leadership in open innovation communities, *Organization Science*, 18(2), pp. 165–180.

Franke, N. and Shah, S. (2003) How communities support innovative activities: an exploration of assistance and sharing among end-users, *Research Policy*, 32(1), pp. 157–178.

Glaser, B. G. and Strauss, A. L. (1967) *The Discovery of Grounded Theory; Strategies for Qualitative Research* (Chicago: Aldine).

Godwin, M. (1994) Nine principles for making virtual communities work, *Wired*, 2.06.

Hargadon, A. B. and Douglas, Y. (2001) When innovations meet institutions: Edison and the design of the electric light, *Administrative Science Quarterly*, 46(3), pp. 476–501.

Hargrave, T. J. and van de Ven, A. H. (2006) A collective action model of institutional innovation, *Academy of Management Review*, 31(4), pp. 864–888.

Hars, A. and Ou, S. (2002) Working for free? Motivations for participating in open-source projects, *International Journal of Electronic Commerce*, 6(3), pp. 25–39.

Hertel, B., Niedner, S. and Herrmann, S. (2003) Motivation of software developers in open source projects: an Internet-based survey of contributors to the Linux kernel, *Research Policy*, 32(7), pp. 1159–1177.

Hunt, C. S. and Aldrich, H. E. (1998) The second ecology: creation and evolution of organizational communities, *Research in Organizational Behavior*, 20, pp. 267–301.

Isaak, J. (2006) The role of individuals and social capital in POSIX standardization, *International Journal of IT Standards and Standardization Research*, 4(1), pp. 1–23.

Jeppesen, L. B. and Frederiksen, L. (2006) Why do users contribute to firm-hosted user communities? The case of computer-controlled music instruments, *Organization Science*, 17(1), pp. 45–63.

Lakhani, K. R. and von Hippel, E. (2003) How open source software works: "free" user-to-user assistance, *Research Policy*, 32(6), pp. 923–943.

Lakhani, K. and Wolf, R. G. (2003) Why hackers do what they do: understanding motivation and effort in free/open source software projects. MIT Sloan Working Paper No. 4425-03, September. Available at http://freesoftware.mit.edu/papers/lakhaniwolf.pdf.

Lee, G. K. and Cole, R. E. (2003) From a firm-based to a community-based model of knowledge creation: the case of the Linux kernel development, *Organization Science*, 14(6), pp. 633–649.

Lerner, J. and Tirole, J. (2002) Some simple economics of open source, *Journal of Industrial Economics*, 52(2), pp. 197–234.

Lerner, J. and Tirole, J. (2005) The scope of open source licensing, *Journal of Law, Economics, and Organization*, 21(1), pp. 20–56.

MacCormack, A. and Herman, K. (1999) Red Hat and the Linux revolution, Harvard Business School Case 9-600-009.

MacCormack, A., Rusnak, J. and Baldwin, C. Y. (2006) Exploring the structure of complex software designs: an empirical study of open source and proprietary code, *Management Science*, 52(7), pp. 1015–1030.

Markus, M. L. (2007) The governance of free/open source software projects: monolithic, multidimensional, or configurational?, *Journal of Management and Governance*, 11(2), pp. 151–163.

Mowery, D. C. and Simcoe, T. (2005) Public and private participation in the development of and governance of the Internet, in: R. R. Nelson (Ed.), *The Limits of Market Organization* (New York: Russell Sage).

Murray, F. and O'Mahony, S. (2007) Exploring the foundations of cumulative innovation: implications for organization science, *Organization Science*, 18(2), pp. 1006–1021.

O'Mahony, S. (2002) The emergence of a new commercial actor: community managed software projects, Unpublished Dissertation, Stanford University.

O'Mahony, S. (2003) Guarding the commons: how community managed software projects protect their work, *Research Policy*, 32(7), pp. 1179–1198.

O'Mahony, S. (2005) Non-profit foundations and their role in community–firm software collaboration, in: J. Feller, B. Fitzgerald, S. A. Hissam & K. R. Lakhani (Eds), *Perspectives on Free and Open Source Software*, pp. 393–413 (Cambridge, MA: MIT Press).

O'Mahony, S. (2007) The governance of open source initiatives: what does it mean to be community managed?, *Journal of Management and Governance*, 11(2), pp. 139–150.

O'Mahony, S. and Ferraro, F. (2007) The emergence of governance in an open source community, *Academy of Management Journal*, 50(5), pp. 1079–1106.

O'Reilly, T. (2005) The open source paradigm shift, in: J. Feller, B. Fitzgerald, S. A. Hissam & K. R. Lakhani (Eds), *Perspectives on Free and Open Source Software*, pp. 461–482 (Cambridge, MA: MIT Press).

Prügl, R. and Schreier, M. (2006) Learning from leading-edge customers at The Sims: opening up the innovation process using toolkits, *R&D Management*, 36(3), pp. 237–250.

Raymond, E. (1999) *The Cathedral and the Bazaar: Musings on Linux and Open Source by an Accidental Revolutionary* (Sebastopol, CA: O'Reilly).

Robbins, J. (2005) Adopting open source software engineering (OSSE) practices by adopting OSSE tools, in: J. Feller, B. Fitzgerald, S. Hissam & K. Lakhani (Eds), *Perspectives on Free and Open Source Software*, pp. 245–264 (Cambridge, MA: MIT Press).

Rosen, L. (2004) *Open Source Licensing: Software Freedom and Intellectual Property Law* (Upper Saddle River, NJ: Prentice Hall).

Rosenkopf, L. and Tushman, M. L. (1994) The coevolution of technology and organization, in: J. Baum & J. Singh (Eds), *Evolutionary Dynamics of Organizations*, pp. 403–424 (New York: Oxford University Press).

Rosenkopf, L. and Tushman, M. L. (1998) The co-evolution of community networks and technology: lessons from the flight simulation industry, *Industrial and Corporate Change*, 7(2), pp. 311–346.

Rosenkopf, L., Metiu, A. and George, V. P. (2001) From the bottom up? Technical committee activity and alliance formation, *Administrative Science Quarterly*, 46(4), pp. 748–772.

Sabel, C. F. (1984) Industrial reorganization and social democracy in Austria, *Industrial Relations*, 23(3), pp. 344–362.

Saxenian, A. (1994) *Regional Advantage: Culture and Competition in Silicon Valley and Route 128* (Cambridge, MA: Harvard University Press).

Schoonhoven, C. B. and Romanelli, E. (Eds) (2001) *The Entrepreneurship Dynamic: Origins of Entrepreneurship and the Evolution of Industries* (Stanford, CA: Stanford University Press).

Shah, S. (2006) Motivation, governance, and the viability of hybrid forms in open source software development, *Management Science*, 52(7), pp. 1000–1014.

Stinchcombe, A. (1965) Social structure and organizations, in: J. G. March (Ed.), *Handbook of Organizations* (Chicago: Rand McNally).

Strauss, A. L. and Corbin, J. (1990) *Basics of Qualitative Research: Grounded Theory Procedures and Techniques* (Newbury Park, CA: Sage).

Tushman, M. and Rosenkopf, L. (1992) Organizational determinants of technological change, *Research in Organizational Behavior*, 14, pp. 311–347.

Välimäki, M. (2003) Dual licensing in open source software industry, *Systemes d'Information et Management*. Available at http://opensource.mit.edu/papers/valimaki.pdf.

van de Ven A., H. (1993) A community perspective on the emergence of innovations, *Journal of Engineering and Technology Management*, 10(1/2), pp. 23–51.

van de Ven, A. and Garud, R. (1989) A framework for understanding the emergence of new industries, *Research on Technological Innovation, Management and Policy*, 4, pp. 195–225.

von Hippel, E. (1988) *The Sources of Innovation* (New York: Oxford University Press).

von Hippel, E. (2001) Innovation by user communities: learning from open-source software, *MIT Sloan Management Review*, 42(4), pp. 82–86.

von Hippel, E. (2005) *Democratizing Innovation* (Cambridge, MA: MIT Press).

von Hippel, E. and von Krogh, G. (2003) Open source software and the "private-collective" innovation model: issues for organization science, *Organization Science*, 14(2), pp. 209–223.

von Krogh, G., Spaeth, S. and Lakhani, K. R. (2003) Community, joining, and specialization in open source software innovation: a case study, *Research Policy*, 32(7), pp. 1217–1241.

West, J. (2003) How open is open enough? Melding proprietary and open source platform strategies, *Research Policy*, 32(7), pp. 1259–1285.

West, J. (2007) Value capture and value networks in open source vendor strategies, *Proceedings of the 40th Annual Hawai'i International Conference on System Sciences*, Waikoloa, Hawai'i, p. 176.

West, J. and Dedrick, J. (2001) Open source standardization: the rise of Linux in the network era, *Knowledge, Technology & Policy*, 14(2), pp. 88–112.

West, J. and Gallagher, S. (2006) Challenges of open innovation: the paradox of firm investment in open-source software, *R&D Management*, 36(3), pp. 319–331.

West, J. and O'Mahony, S. (2005) Contrasting community building in sponsored and community founded open source projects, *Proceedings of the 38th Annual Hawai'i International Conference on System Sciences*, Waikoloa, Hawai'i, p. 196c.

Williams, R. L. and Cothrel, J. (2000) Four smart ways to run online communities, *Sloan Management Review*, 41(4), pp. 81–91.

Keep it Simple: A Companion for Simple Wikipedia?

MATTHIJS DEN BESTEN & JEAN-MICHEL DALLE

ABSTRACT In this paper, we inquire about some of the ways in which the community around Simple Wikipedia—an offspring of Wikipedia, the notorious free online encyclopedia—manages the online collaborative production of reliable knowledge. We focus on how it keeps its collection of articles "simple" and easy to read. We find that the labeling of pages as "unsimple" by core members of the community plays a significant but seemingly insufficient role. We suggest that the nature of this mode of decentralized knowledge production and the structure of Wiki-technology might call for the implementation of an editorial companion to the community.

1. Introduction

Wikipedia is a collection of articles that has come to occupy a prominent place on the World Wide Web (Cuhalev, 2007). Recently, there has been a lively debate on the reliability and quality of this encyclopedia (cf. Denning *et al.*, 2005; Giles, 2005), but we contend that the question how the quality of Wikipedia can be sustained and improved is even more important. Ultimately, we would argue, this sustainability is a question of management. While open science may be a fragile cultural legacy that depends on specific forms of patronage (David, 2004), for open content it would seem sufficient that there are people who are willing to contribute (von Hippel, 2007). Yet, it is important to keep in mind that the establishment and maintenance of a collection of articles does not only require the free flow of contributions, but also considerable effort to edit these contributions, integrate them with the rest of the collection and maintain consistency and scope of the collection as a whole (Mateos Garcia and Steinmueller, 2003). In this respect, Wikipedia is not unlike other complex products and systems (Davies and Hobday, 2007). And it is these organizational capabilities (Dosi *et al.*, 2000) that we would like to study here.

The focus of our study is Simple Wikipedia, an offspring of the main English Wikipedia. Simple Wikipedia was set up as an effort to create a collection of articles that are easy to read for children or non-native speakers of English. Compared to the wider aims of quality and reliability that the main Wikipedia aspires to, readability is a very modest and tractable goal. At the same time, with readability as *raison d'être*, the question whether Simple Wikipedia is a success completely depends on its ability to maintain its readability and so we are likely to see a community that tries its best. As it turns out, maintaining the readability of Simple Wikipedia requires a lot of effort—effort that is often hard to muster. Fortunately, we see a lot of scope for alleviation as artificial companions can be created to help identify spots lacking in readability. More in general, we suggest that partial automation will be necessary to facilitate the tasks of integration and orchestration that tend to be neglected in horizontal, user-only, content networks.

In what follows, we first lay out the rationale for artificial companions in content management. We then give an overview of Simple Wikipedia and point out that readability remains a problem. Finally, we illustrate why, in the case of Simple Wikipedia, management efforts in themselves are not sufficient to overcome the problem and why the construction of artificial companions may be the way forward.

2. Community Companions

Companion is a term coined by Wilks (2005) for software acting as a personal assistant, which accompanies people in their digital activities and helps make sense of the data they encounter. In the context of Wikipedia, what we are looking for is a script, which not only acts as a personal assistant to individual contributors but which also serves as what Schmidt and Simone (1996) call a "coordination mechanism" for the community as a whole. That is, such a *community companion* would not just make the tasks of integration and editing more efficient, it would also help orchestrate which human resources are deployed where.

We feel that companions are needed because the organizational capabilities of the horizontal, user-only, content networks identified by von Hippel (2007) would otherwise be lacking. The reason for this is that horizontal, user-only, content networks probably cannot be completely flat and typically won't consist of just one type of user given what we know: not only do simulations suggest that peer production works best when there is a mix of collaborators with distinct motivations (Dalle and David, 2007), moreover, from a conceptual point of view, content production is an amalgam of multiple roles (Mateos Garcia and Steinmueller, 2003). In their discussion of the various user-roles that are needed for the construction and maintenance of collections and systems, Mateos Garcia and Steinmueller contrast open source systems like Apache with open content collections like Wikipedia and argue that in the latter case contributors will have less incentive to participate in the integration and orchestration of the whole. The reason for this is that while in the case of a software system, the return on individual contributions depends to a large extent on their ability to function effectively as part of a system, in the case of an article collection, individual contributions depend less on others and could be very valuable as self-publications as well. However, if contributors have little incentive to integrate, the danger is that the collection will be less integrated and coherent then is optimal from the point of view of the user-as-reader as well. Here is where firms could step in and pay people to make up for the effort. This is, for instance, the model of Answers.com, which represents content from Wikipedia in an arguably more reader-friendly format. But there is also scope for firms to host their own commercial online communities (Wiertz and de Ruyter, 2007) just as it has been a long established practice for firms to pay moderators in online chat rooms. Yet, in an organization like Wikipedia, which is set up as a non-profit organization, does not want to sell advertisements and intends to keep overheads low, the exploration of more automation may be necessary.

Within Wikipedia, several important steps have already been made towards the creation of a community companion. First of all, to a large extent, the MediaWiki software, which serves as the system through which contributions are added and changes are made to articles in Wikipedia, also enforces a common "look and feel" for the whole collection. In addition, over the years several software scripts have been written and employed to help contributors with their work. Often these scripts scour large sets of data and carry out or suggest edits by themselves and in reflection of this automatic nature, in Wikipedia these scripts are called "bot"—an abbreviation of robot. The first bot appeared in October 2002, not long after Wikipedia was initiated. This bot, developed by Derek Ramsay, automatically generated articles about towns in the USA on the basis of US census data (Wikipedia, 2008). Many more bots followed. Many of them were involved in simple tasks like the establishment of links between Wikipedia articles in different languages, which are typical for an integrator, or automated spell checks more typical for an editor. But bots are also increasingly relied on for more complex tasks: it has been speculated that the widespread adoption of anti-vandalism bots has helped stem the rise of vandalism (Priedhorsky *et al.*, 2007) and a "Suggestbot" has recently been developed that suggests articles to edit on the basis of a contributor profile, his or her edit history and the current state of the collection as a whole (Cosley *et al.*, 2007).

Still, there is room for improvement. Considering the organization of information quality work that emerged in Wikipedia over the years (Stvilia *et al.*, 2008), it appears that Wikipedia is increasingly adopting the attributes of a firm: a layer of middle management has been introduced as a small but steadily growing number of users have been awarded special "administrator" rights that allow them to restrict access to other users and remove complete articles from the collection. In addition, there has been a lot of experimentation with the introduction of managerial activities such as rewarding exceptional contributions by donating "barn-stars" to contributors (Kossinets, 2008) or "featuring" articles (Viégas *et al.*, 2007). Moreover, the tagging of articles with labels such as "stub" for articles that need expanding or, in the case of Simple Wikipedia, "unsimple" for articles whose readability needs to be improved can be interpreted as "directives" expressed by management to coordinate human resources (Penrose, 1959; Turvani, 2001). It is more likely that directives are heeded when they are applied consistently and in line with the priorities of the community. As we shall see for Simple Wikipedia, this is an area where community companions could be of great help.

3. Simple Trends

That all is not well with Wikipedia is illustrated by the case of Simple Wikipedia.

Simple Wikipedia is a spin-off of Wikipedia that was initiated in 2003 because people felt that many articles in Wikipedia were too hard to read, due to the use of jargon, excessive formality or for other reasons—especially for children and non-native speakers of English. Simple Wikipedia, it was hoped, would be the place where people go to look for an easily readable description of topics. At the same time, contributors to Simple Wikipedia would commit to the ideals of this sub-project or at least adhere to an editorial policy that calls for greater readability when contributing descriptions of topics.

What makes Simple Wikipedia a thankful target for analysis is that its success in achieving its goal can be measured to a great extent as various indicators have been developed to rate the readability of texts. In the publishing industry it is common to employ these indicators to guide editors of, for instance, a children's encyclopedia (Bingham, 2007). The Flesch reading easy formula, which we employ for our analysis here, is such an indicator (Flesch, 1979). Its score is determined as follows:

$$\text{Score} = 206.835 - 84.6 * (\text{syllables per words}) - 1.1015 * (\text{words per sentence}).$$

The score thus found is a number between 0 and 100 that corresponds to a readability level ranging from very difficult to very easy as is illustrated in Table 1. Despite the simplicity of the formula, the Flesch score has proved to be an often very accurate approximation of readability in English and it can be included as part of the spell checking in the word processor Word.

Like for many other Wikipedia projects, archives of Simple Wikipedia are maintained at downloads.wikimedia.org. These archives contain the raw text sources of all versions of all pages in Simple Wikipedia. On the basis of this information, it is possible to reconstruct the history of Simple Wikipedia: for every month that Simple Wikipedia existed, the archives allow us to retrieve the versions of the articles at that point in time in so far as they existed. Thus, we can reconstruct the size of the collection in that month and compute the overall readability of that collection. Figure 1 shows the result of such a reconstruction for Simple Wikipedia from January 2003 until December 2006. The figure shows an upward sloping line indicating the total number of articles in the collection for each month and a downward sloping line indicating the overall readability of the articles in the collection for each month. Note that the continuous growth in the number of articles in Simple Wikipedia is at least partially due to the fact that articles that have been removed from the collection and are not

Table 1. Interpretation of Flesch scores on texts in English

Flesch reading ease score	Readability level
0–29	Very difficult
30–49	Difficult
50–59	Fairly difficult
60–69	Standard
70–79	Fairly easy
80–89	Easy
90–100	Very easy

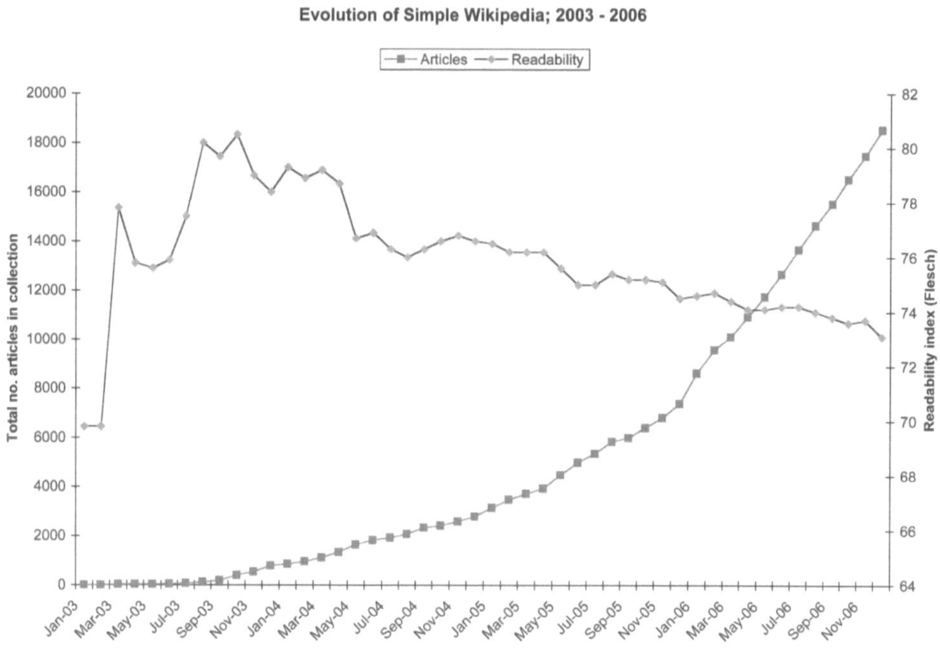

Figure 1. Readability trend and collection size of Simple Wikipedia

currently available anymore, do not appear in the most recent archive either. Even so, the growth in the number of articles over time is impressive. This rate of growth may be a factor that explains why the readability indicated by the Flesch readability score shows a gradual decline: one interpretation that suggests itself is that as the number of articles in Simple Wikipedia grew, it became more unwieldy and editors faced an ever harder task to maintain the standards of readability. With a readability index of over 70, Simple Wikipedia still scores very well given that Standard English falls between 60 and 70. Still, the fact that readability continues to decline should be worrying.

4. "Unsimple" Directives

To stem the decline in readability in Simple Wikipedia, its editors have tried a variety of methods. The most visible of these methods was to tag articles that are considered hard to read with the label "unsimple". The effect of this is to send out an open invitation to whoever comes across the article to fix the article. Moreover, all offending articles are also listed on a separate page that may be consulted by people who are particularly interested in improving readability. Thus the tagging of an article can be said to initiate "broadcast search" (Jeppesen and Lakhani, 2006) by soliciting help to fix an identified problem.

The potential of this mechanism is neatly illustrated in Figure 2. This figure depicts the exemplary evolution of the readability over time of page 3010—a page aptly entitled "Propaganda". Readability is based on the Flesch reading easy score and time is given as the number of days relative to the first occurrence of the label "unsimple" on the page. In

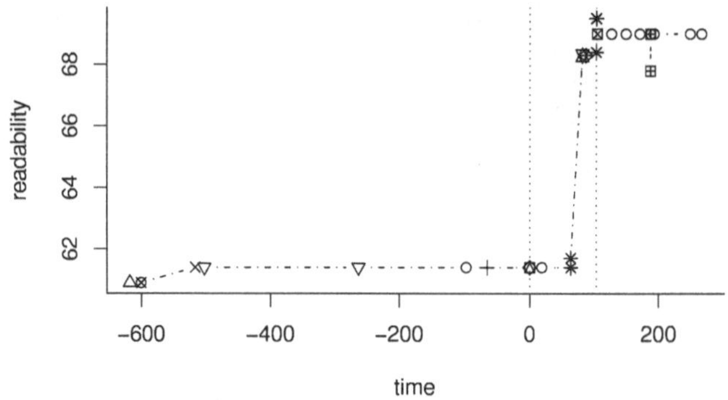

Figure 2. Edits on page 3010—"Propaganda"

addition, the dotted vertical lines indicate the period during which the page was labeled with the label "unsimple" and the shapes of the points correspond to the different contributors to the page. The first thing to notice is the dramatic effect that the label "unsimple" seems to have on the readability of the page. Before the label first appears, the readability score of the page hovers around the border between a standard and a fairly difficult level of readability. When the label is removed, 100 days later, readability has jumped to the other side of the spectrum, now bordering between a standard and a fairly easy level of readability. Looking at the shapes of the points, a history emerges in which someone wrote the initial page to which others then added a bit without changing much in terms of readability. At some point, day 0, a new contributor comes along and points out that the page is too difficult to read for an article in the Simple Wikipedia collection. The label "unsimple" then attracts the attention of someone else who actively starts working to improve the readability of the page. As soon as the problem has been remedied, the label "unsimple" is removed and regular maintenance is resumed.

When we look at the distribution of readability scores of all pages that have had the label "unsimple" applied and removed during their existence for pages from the Simple Wikipedia archive of 6 September 2007, on average, these pages have a score of 61.10 when they are first tagged and a score of 70.67 when the tag is removed, while the average score of all pages in the collection is 69.30. Yet, these are average scores and the variance in the distribution of scores is such that it seems more likely that the judgment whether a page is sufficiently simple was based on the participants' intuition. In that case, the close correspondence of the average scores to the outer ends of the readability spectrum of Standard English merely corroborates the Flesch reading easy formula as a measure for readability. In any case, *t*-tests on the distributions of reading scores reveal that the scores of pages that are tagged are significantly different from the population as a whole when the "unsimple" label is applied and that these scores are indistinguishable from the population as a whole at the point when the label is removed. Furthermore, a one-sided paired *t*-test comparing pages before and after tagging yields a significant improvement in readability—not only for the readability represented by the Flesch reading easy formula, but also for the Kincaid formula, the Automated Readability Index, the Coleman Liau formula, the Fog index, the Lix formula and SMOG-grading. With regards to the speed with which "unsimple" pages are transformed into simple ones, we find an average improvement of almost a half point (0.38) with a confidence of over 95 percent already after 1 day and after 1 week the improvement has become almost a full point (0.90) with a confidence of over 99 percent. Note however that the sample of pages that are affected is relatively small: while the archive contains information about almost 30,000 pages of which almost 20,000 concern articles, there are only 251 pages that have been tagged as "unsimple" at least once in their lifetime. Of these 251 pages, 101, that is, about two-fifths, had the tag removed at a later stage. In a small number of cases, the Flesch score of a page cannot be determined—for instance, when the page has no content—and so there were 245 pages on which we could carry out a paired *t*-test after 1 day, 242 after 1 week and 94 between the first and the last occurrence of the tag.

However, the process of labeling or tagging does not always work. Figure 3 shows examples of pages that exhibit common problems. Again the evolution of the readability over time is shown using the Flesch readability formula and time measured in days. The point type indicates the status of the pages, "+" when the label "unsimple" is absent and "○" when it is there. The titles above the plots correspond to the page identifier and page title in the Simple Wikipedia archive. In some cases, like for the pages on "Earth" and "Steel" the label is applied to a page that is already fairly easy to read. In the case of "Earth", after a considerable delay, the readability increases still further, but in case of "Steel", the presence of the label seems to have provoked a rather adverse reaction with the page evolving from "very easy" to "fairly difficult". In other cases, like for the pages on "M-theory" and "FreeBSD", the label seems to have had the desired effect, but no one has bothered to remove the label afterwards. In a case like for the page on "Neutrino", the label didn't have any effect the first time around, but was noticed and acted upon the second time. However, in a case like the page on "Vagina", an improvement in readability while it was tagged "unsimple" turned into a rapid deterioration once this reminder wasn't there anymore. Another case where the label seems to have been removed too early is "Cold-fusion" as it undergoes rapid improvement while it is officially "unsimple" but does not continue this process when the label is removed again and remains "fairly difficult" to understand.

Erratic behavior in specific cases is not the only problem that hampers the process of labeling. Another problem is that not all badly readable pages are detected. Of the 27,246 pages in our archive that have never been labeled as "unsimple", 2,003 (7 percent) have a maximum Flesch readability score of less than 60 and 897 (3 percent) have a maximum of less than 50. Most of these pages are probably difficult to read. Conversely, there are cases in which the application of a label is not necessary: there are 3,631 pages that evolved from a readability score of less then 60 to a score of more than 70 without ever being labeled.

5. Simple Companions

Many of the problems with the "unsimple" directive identified above would actually be quite easy to address. The main problem seems to be that there are only a small number of people who engage in tagging. Besides, these people often fail to follow up and remove the tag once an article has become much more readable and sometimes their judgment of readability seems to be a bit off the mark. A solution could be to develop a program that scans articles for readability and suggests tagging with "unsimple" when a score is very low. Alternatively, the program could identify articles that are tagged as "unsimple" but seem

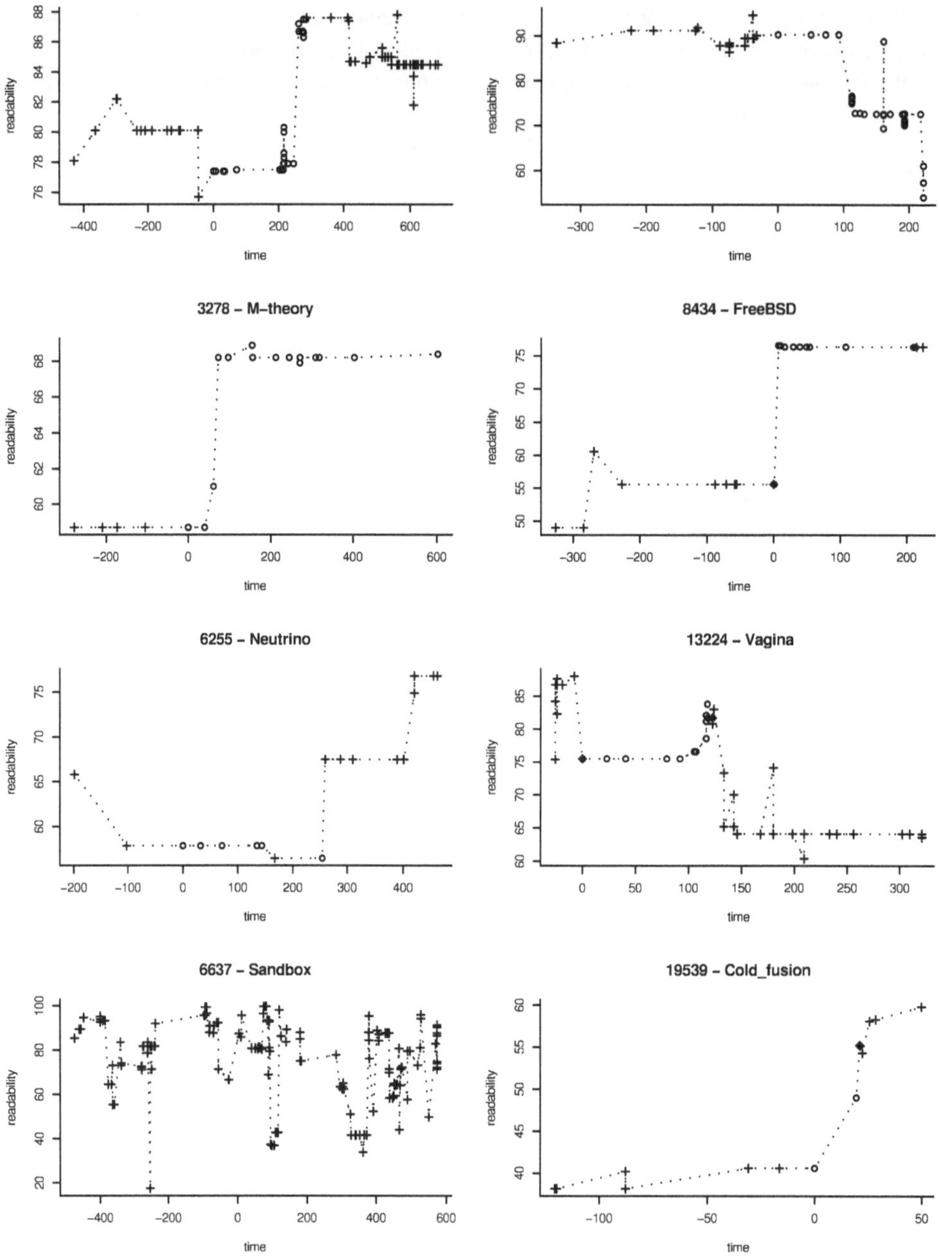

Figure 3. Keeping it simple: a selection of pages

perfectly readable or point towards articles that have been labeled "unsimple" where the readability score has not changed a lot over a long time. Yet, we don't want to suggest that the Flesch reading easy score is the be-all and end-all of readability. One could consider different indicators. For instance, the program could check for vocabulary and allow specific words while disallowing others. As long as that program is available to all editors of Simple Wikipedia and more or less reflects the goals of this collection, it would do a good job facilitating the maintenance tasks of these editors and help sustain Simple Wikipedia in a consistent way.

Implementing a companion would also be a relatively simple solution for Wikipedia as a whole in that it would both respect its characteristic decentralized features while harnessing the creative work of volunteers one step further: beyond the various low-level scripts or bots that now exist, to an open source and more fully integrated companion whose design and implementation would be the result of the cumulative and decentralized efforts of the community. Such a community companion would not only measure readability, but would also keep track of various other indices of quality like the ones listed by Stvilia *et al.* (2005). Many of the "directives" appearing in Wikipedia nowadays, be it "barn-stars", "stub" and "controversy" labels, or "featured article" awards denote conditions that can be measured in one way or the other. Having a community companion would promote the consistency of those "directives" and at the same time it would also be more transparent why they appear, which, one would hope, would promote the effectiveness of these "directives". For horizontal, user-only, content networks in general, the promise of companions is that it can automate or at least alleviate many of the tasks that would not normally be carried out by users themselves, thus increasing the viability of these networks.

6. Conclusion

We believe that the management and maintenance of collaborative knowledge production is soon to become a very important but difficult issue. The practice to identify pages that need to be attended seems to work although many pages go undetected, and while in some cases the tag that calls for attention is ignored or could even provoke an adverse reaction. However, on the positive side, we find that relatively simple metrics could go a relatively long way to help the community analyze and monitor the process of tagging. This suggests that it would be feasible to build appropriate bots, and perhaps a community companion as an editorial assistant for the management and maintenance of collections of this kind.

Acknowledgements

A preliminary version of this paper was accepted for and presented to the EURAM 2007 Conference held in Paris, 16–19 May 2007. The authors thank the participants and three anonymous reviewers for their comments and suggestions. They also thank the reviewers and editors of *Industry and Innovation* for their subsequent comments and suggestions.

References

Bingham, J. (2007) Personal communication, 11 July.

Cosley, D., Frankowski, D., Terveen, L. and Riedl, J. (2007) Suggestbot: using intelligent task routing to help people find work in Wikipedia, *Proceedings of the 12th International Conference on Intelligent User Interfaces*, Honolulu, Hawaii, pp. 32–41 (ACM).

Cuhalev, J. (2007) Ranking of Wikipedia articles on search engines for searches about its own articles. Available at http://www.kiberpipa.org/gandalf/blog-files/wikistatus/wikistatus.pdf (accessed 30 November 2007).

Dalle, J.-M. and David, P. A. (2007) Simulating code growth in libre (open-source) mode, in: N. Curien & E. Brousseau (Eds), *The Economics of the Internet* (Cambridge: Cambridge University Press).

David, P. A. (2004) Understanding the emergence of "open science" institutions: functionalist economics in historical context, *Industrial and Corporate Change*, 13, pp. 571–589.

Davies, A. and Hobday, M. (2007) *The Business of Projects: Managing Innovation in Complex Products and Systems* (Cambridge: Cambridge University Press).

Denning, P., Horning, J., Parnas, D. and Weinstein, L. (2005) Wikipedia risks, *CACM*, 12(48).

Dosi, G., Nelson, R. and Winter, S. (Eds) (2000) *The Nature and Dynamics of Organizational Capabilities* (Oxford: Oxford University Press).

Flesch, R. (1979) *How to Write Plain English* (New York: Harper and Row).

Giles, J. (2005) Internet encyclopaedias go head to head, *Nature*, 438, pp. 900–901.

Jeppesen, L. B. and Lakhani, K. R. (2006) Broadcast search in problem solving: attracting solutions from peripheral solvers. Technical Report 2006-10-30, Harvard Business School.

Kossinets, G. (2008) Coordination and self-organization in Wikipedia. Presentation at Nuffield College, Oxford, January.

Mateos Garcia, J. and Steinmueller, W. E. (2003) Applying the open source development model to knowledge work. INK Open Source Research Working Paper 2, SPRU—Science and Technology Policy Research, University of Sussex, January.

Penrose, E. (1959) *The Theory of the Growth of the Firm* (Oxford: Basil Blackwell).

Priedhorsky, R., Chen, J., Lam, S. K., Panciera, K., Terveen, L. and Riedl, J. (2007) Creating, destroying, and restoring value in Wikipedia, in: T. Gross & K. Inkpen (Eds), *GROUP'07*.

Schmidt, K. and Simone, C. (1996) Coordination mechanisms: towards a conceptual foundation of CSCW systems design, *Computer Supported Cooperative Work (CSCW)*, 5(2), pp. 155–200.

Stvilia, B., Twidale, M. B., Smith, L. C. and Gasser, L. (2005) Assessing information quality of a community-based encyclopedia, *Proceedings of the International Conference on Information Quality*.

Stvilia, B., Twidale, M. B., Smith, L. C. and Gasser, L. (2008) Information quality work organization in Wikipedia, *Journal of the American Society for Information Science and Technology*, 59(6), pp. 983–1001.

Turvani, M. (2001) Microfoundations of knowledge dynamics within the firm, *Industry and Innovation*, 8(3), pp. 309–323.

Viégas, F. B., Wattenberg, M. and McKeon, M. M. (2007) The hidden order of Wikipedia, in: D. Schuler (Ed.), *Online Communities and Social Computing*, Vol. 4564 of LNCS, pp. 445–454.

von Hippel, E. (2007) Horizontal innovation networks—by and for users, *Industrial and Corporate Change*, 16, pp. 293–315.

Wiertz, C. and de Ruyter, K. (2007) Beyond the call of duty: why customers contribute to firm-hosted commercial online communities, *Organization Studies*, 28(3), pp. 347–376.

Wikipedia, History of Wikipedia bots. Available at en.wikipedia.org/wiki/Wikipedia: History_of_Wikipedia_bots (accessed 20 January 2008).

Wilks, Y. (2005) Artificial companions, in: *Machine Learning for Multimodal Interaction*, LNCS, pp. 36–45 (Berlin: Springer).

Communities of Consumption and Made in Italy

ELEONORA DI MARIA & VLADI FINOTTO

ABSTRACT The interest towards the role of user communities in innovation has grown among scholars and practitioners. Research has explored the role of communities in high-tech and medium-tech industries with a focus on innovation in the functional dimension of products. Less attention has been devoted to user communities' contribution in industries such as fashion, where innovation is much more related to communication and aesthetics. This paper provides a preliminary set of concepts and working hypotheses regarding the contribution of communities to the non-functional dimension of product innovation in low-tech industries and to the relationship between user involvement in brand communities and their incentives to contribute to innovation both tangible and intangible. The paper discusses two case studies of Made in Italy enterprises that refer to communities for their innovation strategies.

1. Introduction

The interest towards the role of user communities in innovation processes has grown among scholars and practitioners. The need to know more about the contribution of user communities in innovation processes stems from the increased visibility of these collective actors in several industries—from software to sporting equipment and many others. It has been widely shown that many consumers are far from being passive recipients of products and technical solutions developed by companies. They are rather active contributors in the process of new product development, providing valuable and useful technical knowledge, which can be effectively deployed in innovation processes (Nambisan, 2002; von Hippel and von Krogh, 2003). What determined the rise in the interest towards user intervention in innovation in the last 10 years is the scale of distributed innovation processes participated in by users that information and communication technologies (ICT) have made possible (Hagel and Armstrong, 1998; von Hippel, 2005). Web and collaborative technologies allow individuals scattered in distant geographical areas to gather in specific online websites and collaboration areas and allow for the enactment of a potentially global pool of intelligence and capabilities (Sproull and Kiesler, 1991; Rheingold, 2002).

This paper discusses the contribution of consumer communities to the non-functional dimension of product innovation in low-tech industries, by focusing on Made in Italy small and medium-sized enterprises (SMEs). Our paper intends to offer an original point of view that links innovation literature with marketing and consumer research approaches, traditionally not adopted in the analysis, but fundamental to understand the role of user communities in innovation. From our perspective, innovation entails both the tangible and the intangible dimensions of product for customers' value creation and firm competitiveness. Innovation processes blend inputs from customers both in terms of technical knowledge rooted in the contexts of use and shared meanings and identities, exploited in firms' brand strategies (Muniz and O'Guinn, 2001).

Research in innovation studies has addressed the role of consumer communities especially in high-tech industries, in particular in the software and information and communication industries. Extant research and literature have focused diffusely on user contribution to functional and technical innovation (von Hippel, 1988). Instead, less attention has been devoted to user communities' contribution to innovation in industries such as fashion. Traditionally, the economic and management literature have defined these industries as "mature" or low-tech industries, by highlighting how innovation in fashion, for example, is much more related to communication and aesthetics than to functional features (i.e. Ravasi and Lojacono, 2005).

As many studies in this field have highlighted, innovation in traditional industries insists eminently on the intangible dimensions of products, rather than solely on their functionality (Krippendorf and Butter, 1984; Verganti, 2006). From this standpoint, the creation of experiences for customers is a fundamental part of firms' strategies in the mature industries to generate value (Schmitt and Simonson, 1997; Pine and Gilmore, 1999, 2007; Prahalad and Ramaswamy, 2003). Our perspective is that understanding and managing innovation in the present competitive scenario requires a stronger integration between innovation studies—traditionally focused on R&D and technological innovation in product development—and the analyses offered by the marketing and customer behavior framework (i.e. Veryzer, 2005). Experiential marketing, relationship marketing approaches and studies on services stress the role of customers in value creation, where customers and the firm co-create the firm's offerings (i.e. Normann, 2001; Lush *et al.*, 2007).

In this approach, consumer communities can play an active role in such innovation processes. However, the nature of their contribution and the dynamics underlying it have been only partially described and formalized, as studies mainly focused on the functional side of products and gave minor emphasis on the communicative and aesthetic aspects of innovation. Literature on brand communities provides an initial set of categories and concepts (e.g. Muniz and O'Guinn, 2001) useful to fill this gap. Brand communities can sustain customers' involvement in the sense-making process—as a key component of innovation—which is activated by the firm but requires also customers' participation.

As successful cases show—that is, the example of Apple (*The Economist*, 2007)—innovation is not always limited to technical and technological creativity (functional dimension of products), but it also or mainly covers artistic/aesthetic creativity (forms and meanings associated to products and brands). In our perspective innovation is the result of both these processes, where the role of communities of customers is related to support innovation from multiple directions and the role of consumption assumes a greater relevance on both sides (Bettiol and Micelli, 2005). As other studies pointed out (i.e. Sawhney and Prandelli, 2000), communities can be at the core of the innovation process, by shaping and being embedded in the locus of innovation. From our point of view, it is important to consider it also as a locus of experience where meanings and knowledge are developed and shared, among consumers and with companies (Prahalad and Ramaswamy, 2003).

The analysis presented in the paper refers to the Made in Italy economic system based on SMEs operating in mature industries. This specific focus allows us to address specifically consumer communities' contributions to the intangible and non-functional dimension of products, since a vast majority of Italian companies are focused on mature industries such as fashion. Despite many studies analyzing innovation within Italian industrial districts and firms' adoption of network technologies (e.g. Chiarvesio *et al.*, 2004), very few discuss the role of communities of customers in innovation processes of Made in Italy and link those issues with brand strategy, which is relevant for firms' competitiveness in those industries (Corbellini and Saviolo, 2004).

The paper is organized as follows: first, we provide a review of the studies exploring the role of consumption in innovation processes. Second, we analyze the role of customers and user communities in non-functional innovation processes drawing from the marketing and consumer research literature. Third, the paper focuses on the chances for firms in exploiting communities of customers to sustain their competitive advantage through a broader innovation perspective. Next, according to a qualitative methodology, we discuss two selected exploratory case studies of SMEs specializing in Made in Italy industries that refer to communities as relevant elements of their innovation strategies.

2. Online Consumer Communities and Innovation

A large number of studies have shown that users of products and processes often develop fundamental innovations and that relevant learning about new product development happens outside the boundaries of the firm (Chesbrough, 2003; Franke and Shah, 2003; Jeppesen and Molin, 2003; Rosenberg, 1982; von Hippel, 2005). So far, the analyses of user innovation communities have successfully provided evidence regarding the motivations and rewards driving users in their efforts to improve products and processes and in revealing their knowledge freely (Harhoff *et al.*, 2003). In addition, studies in the software industry and in the sporting equipment industry showed that innovation by users is more likely to take place in community settings rather than being a single innovator's solitary process (von Hippel, 2001; von Hippel and von Krogh, 2003; Jeppesen and Fredriksen, 2006).

Many innovations in the sporting equipment industry, for example, have been developed by users, rather than by manufacturers, as documented in Franke and Shah (2003). In the software industry, open source software such as Linux gained a prominent position as an alternative to proprietary software, thanks to the combined effect of user innovation and the active endorsement and contribution by hardware, software and consulting companies. In the electronic games industry, manufacturers are responsible for the creation of the main logical and functional architecture of a game and for the original storylines. After games are commercialized, further innovations in the game play are developed by passionate users, who contribute lines of code, graphics and modeling to the original products improving them and making them more appropriate to peculiar patterns of use and playing experiences (Jeppesen and Molin, 2003).

As stated in von Hippel (1988), the incentives to innovate are different for manufacturers and users. Manufacturers expect to benefit from a given innovation by selling it. On the one hand, this means that what drives the design of a new product by manufacturers is the forecast of a large demand for that product. On the other hand, innovations are likely to be developed by users if they are to benefit from their use (Franke and Shah, 2003). While not all users play such a part in the innovation process, research on the sources of innovation identified a particular type of user able to identify specific and sophisticated needs and to elaborate innovative responses: lead users (von Hippel, 1988, 2005; Morrison *et al.*, 2004). As defined in von Hippel (1988), they are those users that have needs that anticipate general demand in the marketplace and that expect to obtain high benefit from a solution to their needs. The shift of the locus of innovation towards users has been explained as a consequence of information stickiness and its embeddedness in specific contexts of use (von Hippel, 1994). It has also been noted that attributes of the information holder and of his context play a crucial role: users often use languages that are extremely different from those used by manufacturers. Where manufacturers think in terms of processes, technologies and features, users think in terms of performance, practices and desires (Micelli, 2000).

Given the ability of lead users to develop new product concepts, features and prototypes, it is possible for manufacturers to shift part of the innovation process to them. In the software industry, for example, empowerment of customers led a number of software development companies to rely on the competences and skills of users in order to customize their products, responding thus to the needs of customization without having to bear the cost of doing it directly and the risk of not finding adequate markets (Franke and von Hippel, 2003; Jeppesen and Molin, 2003). Sophisticated users needing to adapt a particular application to their local needs can do it through software development kits.

User toolkits for innovation and customization (Franke and Piller, 2004; von Hippel, 2005) have become a major feature in the software as well as in other industries: they allow manufacturers not only to shift part of the burden for customization and innovation to users, but also to exploit commercially user-made innovations if they prove to be successful through user-to-user diffusion.

3. User Communities in Marketing and Consumer Studies: Emerging Perspectives

A second stream of literature focused on consumer communities does not have directly an interest in product innovation processes, but rather it analyzes how communities of consumers are active contributors to the construction of social identities, relational patterns, use contexts within which products are located and within which they are attributed with meaning (Cova, 1997; Hagel and Armstrong, 1998; Kozinets, 1999). Brands in particular have been at the center of the analyses of these works. Much of this literature has focused on the social nature of brands as the result of processes in which companies create complex systems of meanings drawing from particular cultures of consumption (e.g. Cova, 1997).

The creation of meanings and identities qualifying brands does not happen in a social void. Communities of customers represent an important actor in the creation and development of part of brands' values and communicative potential. The process is iterative. On the one hand, brands draw from cultures and subcultures of consumption (Schouten and McAlexander, 1995; Muniz and O'Guinn, 2001). On the other hand, consumers enact these meanings and identities in their everyday consumption and social practices, enriching and sometimes innovating their structure and re-defining the contents of brands. The appropriation of brands and products as signs by consumers aims at communicating their identities and values to other consumers in their everyday life.

These identities and the cultures associated with them are not the result of a singular and individual signification process, but are closely connected to the interaction within groups (Jenkins, 2006). Cova (1997) finds several occurrences of this type of bottom-up identity-building processes: the ravers culture in the UK during the 1990s, the skater culture in France. Being affiliated with these groups means adopting the communication and expressive instruments and semantics created by the group and at the same time contributing to their transformation and evolution through the negotiation of new meanings and new stimuli coming from the experiences consumers have. Cova, drawing from the sociological work of Maffesoli (1992), defines these collective sense-making actors as *tribes*, while many works in the consumer research field and in the marketing field define them as "communities". Within this framework products and brands perform a connecting function, linking consumers and enabling social relations based on distinctive consumption practices. This linking function is a crucial driver of value for customers in many markets for consumer goods, representing a fundamental share of the overall value perceived together with the functional content of goods and services.

3.1. Brand Communities and Collective Sense-making in Consumption as a Form of Innovation

Research on the communicative nature of brands evolved from the analysis of brands as shortcuts for product identification towards the analysis of brands as constructs linking consumers and companies. As stated earlier branding practices are not based on a monologue metaphor (Aaker and Joachimsthaler, 2000), which means they are not the result of a mono-directional communication and signification act by marketers. Designing the communicative arsenal of a brand with pre-defined values and meanings and distributing it to the market through traditional and new media seems a strategy destined to fail. The reason is that being used by consumers as signs for self-expression, brands and products have to reflect particular views of the world and lifestyles, which are pre-existent in consumption contexts and cultures. Works by Belk (1988) and Ahuvia (2005) clearly show that the relationship between consumers and brands—or objects—is determined by the inherent content of the brand itself: as long as it synthesizes the cultures and the identities of specific groups of consumers, they gain in perceived authenticity and become "loved objects". This can happen, though, only if brands reflect the results of sense-making processes taking place in the relational contexts within which consumers are inserted

(Morling and Strannegard, 2004; Roberts, 2004). Rather than being pre-determined and immutable intangible artifacts, brands are narratives that include and continuously re-interpret the transformation that takes place in the consumer arena.

Thus, in many industries, consumers are actually active contributors to the brand: they are responsible for sense-making and identity building through social relationships and shared practices and interests. Is this process good for companies and partly controllable? The literature on consumer communities and particularly brand communities represents a first important step towards the understanding of these dynamics. Schouten and McAlexander (1995) have pointed out the potential impacts that subcultures of consumption can have on marketing practices and strategies. Hard-core members of a specific subculture gain a specific status of opinion leader and often contribute to create subcultural styles of consumption that can be shared and imitated by other consumers and can be detached from those designed by companies.

Their article conducts an in-depth analysis of a particular subculture, that of the "new bikers" who use Harley Davidson motorcycles. The involvement of consumers in the activities and in the life of the subculture generated what the authors call "grassroots R&D": the more involved and technically savvy consumers in the subculture designed and built custom motorcycles which then the company refined and commercialized, such as the chopper and many more. The subculture started also to design and produce their own lines of clothing and aesthetic instruments of expression, giving further marketing ideas and instruments to the company. The user community becomes an active player in the innovation process, partly because of its inputs in terms of product features and partly because it is able to nurture discussion about the products that generate shared meanings (the symbolic dimension of product).

Muniz and O'Guinn (2001) have highlighted the active role played by brand communities, composed by individuals sharing a preference for a specific brand or product (see also Bagozzi and Dholakia, 2006). Passionate consumers of specific brands display an active role in interpreting and re-defining the meanings associated with brands and products, inserting them in a complex context composed by identities, values and other products which are socially negotiated. This active interpretive function of brand communities is responsible for the definition of the intangible value of products and brands and represents a strategic lever for companies willing to incorporate the outcomes of these processes in their offer. Although not specifically addressing the topic of innovation in consolidated terms, literature in consumer studies and marketing contributes to enlarge the definition of innovation itself, highlighting how the non-functional dimension of products—the values and identities associated to brands and goods, the communicative and aesthetic content—constitutes a relevant share of products' value in consumer goods. As noted by Ravasi and Lojacono (2005):

> Even companies that used to compete primarily on price or technological performance are increasingly facing the need to develop products that are not just efficient or reliable, but also responsive to emerging lifestyles and cultural values. (Ravasi and Lojacono, 2005: 52)

In addition, this stream of literature highlights the active role users play in contributing to the definition and the assessment of these elements.

4. Communities and Firms: Innovation Based on Use, Participation and Experience

The contributions of consumer communities to the creation of value, based on the two streams of literature reviewed earlier, can be considered twofold: (i) user communities contribute to the creation of novel technical solutions and functional improvements of products of choice capitalizing their contextual knowledge and their consumption practices; (ii) user communities provide brands and products with part of the symbolic and communicative elements which are responsible for product value.

The issue at stake is not only a theoretical categorization of the different contributions of consumer communities to innovation processes, but rather the nature of the innovation process itself and the value of its outcome. As shown in a growing body of literature in innovation studies, design studies and marketing, innovation is a multidimensional phenomenon, and consumers' perception of the value of a new product depends on two different dimensions: the performance and technical dimension of a product, and its experiential and emotional content, which is embodied in its aesthetics, its style and the values and identities attached to its brands (Krippendorf and Butter, 1984; Pine and Gilmore, 1999). In this perspective, the contributions of marketing and design research are of an extraordinary importance to understand the role of communities of customers in the innovation process.

Today's consumers are increasingly facing a constant flow of innovative functions and features in many of the products they use. Moreover, in many consumer markets alternative products present many similarities under the standpoint of manufacturing excellence and technological content. Based on this evidence, it has been observed that manufacturing activities and manufacturing excellence are not the only determinants of the value perceived by customers (Porter, 1985; Rullani, 2004; Nussbaum *et al.*, 2005).

On the contrary, scholars stated that the determinants at the basis of customers' perceived value resides increasingly, even though not exclusively, in intangible attributes, whose nature has been addressed through several concepts: aesthetics, communication and signification (Hirschman and Holbrook, 1982), luxury (Silverstein and Fiske, 2003), style (Postrel, 2004) and the like. The intangible content of products is said to be the fundamental determinant of consumers' willingness to pay premium prices for products they use, while the lack thereof is responsible for the trading down of goods and services (Silverstein and Fiske, 2003). Silverstein and Fiske note how a general trend is becoming visible in contemporary advanced societies: consumers are willing to pay significant premium prices for goods and services that are emotionally important for them.

These emerging trends do not imply that the inherent quality of products (intended as the outcome of sophisticated manufacturing skills and processes) or their technological and functional content do not play a part in the determination of value. They just re-allocate the relative contribution of tangible and intangible content in the generation of value.

The explanation often brought to these dynamics is that manufacturing capabilities and expertise are becoming commoditized, especially in low-tech industries. Consumers thus perceive that many if not all companies competing for their attention are equivalent in terms of manufacturing capabilities and quality is taken as granted and as the minimum common denominator of all the alternative offerings entering their decision processes (Pine and Gilmore, 2007). In this sense, investing in brands becomes a key strategy for firms to support competitiveness through a wider approach to innovation. As far as technology and innovation are concerned, the marginal importance of technological advancements in markets for consumer goods in generating value for the customer has decreased as a result of the continuous introduction of new products and new features in existing products.

This evidence seems to suggest the need for a re-definition of the concepts and practices of product innovation. Rather than focusing on technologies and continuous technological improvements, firms must gather information from consumers in order to understand their expectations and their inner and often tacit needs (Leonard and Rayport, 2000) and, most of all, they must work on the framing of technological developments and of product re-definitions within a complex discourse matching the requests for meaning and expression coming from consumers (Kotro and Pantzar, 2002). Aesthetic and intangible attributes of products trigger processes of identification which ease consumers' decision-making processes and which attribute products with meanings and transform them in expression instruments, beyond their functional features and quality.

These trends interest particularly mature and low-tech industries, such as fashion, since their products, which have reached a technological *plateau*, are increasingly being differentiated through intangible elements. Innovating means most of all qualifying products through aesthetics and designing complex cultural and communication activities and strategies centered on evocative and emotionally rich brands. Recently, analogous dynamics are also interesting markets for high-tech or medium-tech consumer products, such as computers, digital personal devices (music and movie players), mobile phones, cars and many more (Nussbaum *et al.*, 2005).

Hence, firms should develop explicit processes to connect to lead users and communities of customers as part of their innovation strategies. As far as the nature of the relationship between companies and brand communities is concerned, it can be the result of at least two different strategies. On the one hand, companies can foster the growth of branded communities, which means of groups of people specifically gathered in branded environments—be they the company's website or sponsored events. On the other hand, companies can choose to relate to existing communities that share interests for specific consumption practices, characterized by a strong collective identity and by the mobilization of participants on the basis of these common elements. These communities are not directly connected with one specific brand, but rather consider some brands as being part of their culture and important elements in the definition of their identity. Such user communities gather in specific websites and online environments as well as in offline events, which can be sponsored but which are managed and run by members of the community rather than by personnel within the company. Monitoring, interacting and relating to these communities, even though not on an exclusive basis and within proprietary environments, can be strategic for companies to legitimize their brands and capture valuable suggestions and insights related to potential evolutions of their offerings, and of the specific demand they are addressing (Hagel and Armstrong, 1998; Rémy and Kopel, 2002).

5. User Communities and Made in Italy: Case Studies

The previous considerations regarding the nature of the contribution of consumer communities to product innovation and, more generally, to the creation of value, lead to the question how companies can start and consolidate relationships with these collective actors and how they can leverage these contributions in their business practices and innovation processes. In order to provide preliminary evidence that could stimulate further studies, two cases were selected within the so-called Made in Italy companies, which is the definition given to typical Italian SMEs located in industrial districts operating in low-tech or medium-tech industries. The two case studies analyzed in the following pages represent the preliminary result of a series of observations which have been undertaken in several Made in Italy companies. The case studies presented are exploratory in their objectives (Yin, 1994) and were developed in order to shed light on emerging processes of community involvement in innovation in industries such as fashion and sporting equipment. Rather than providing evidence regarding causal relationships, these cases are meant to provide an illustration of the types of conceptual proposals made in the previous parts of the paper (Siggelkow, 2007), and to stimulate further refinements.

These case studies were selected among a series of analyses of community involvement by Italian companies in traditional Made in Italy industries. Case studies were constructed through several interviews with R&D managers and marketing managers in the interested companies (Di Maria and Micelli, 2007) and through the observation of the companies' online and offline communities—related activity. The selected case studies aim at providing a preliminary analysis of some relevant experiences and a first set of evidence to be developed in further and more structured research efforts.

5.1. Innovation in Made in Italy

Italian production has received great attention over the past decades, due to its strong combination of quality, variety and convenience. Italy's share of "G7" exports is about 66 percent in footwear, 55 percent in leather, 54 percent in gold jewelry and 37 percent in furniture (*source*: UNCTAD). Italy has shown extraordinary levels of export capacity over the years, counterbalanced by strong embeddedness of firms and manufacturing processes in local areas—the industrial districts (Pyke *et al.*, 1990).

Many of the studies on industrial districts link the competitive force of such a form of local manufacturing organization to its internal cohesion (e.g. Pyke *et al.*, 1990). The focus on manufacturing activities stresses the relevance of production processes and related skills and competencies rooted in the local context as the drivers for the competitiveness of the district firms (Becattini and Rullani, 1996). Industrial districts were able to develop profitable innovative processes, strictly related to their industry specialization and their peculiarity (Lazerson and Lorenzoni, 1999). By different degrees of opening of their core processes, firms share manufacturing activities (components etc.) with partners in a collaborative environment, which stimulate reciprocal learning and product improvements.

Italian industrial districts and local supply chains have represented for a long time one of the main strengths of the national industrial systems. Geographical, cultural and linguistic contiguity facilitated the birth and growth of inter-firm networks sustained by trust which allowed for the fragmentation of production cycles, for a great degree of specialization in single phases of production processes and for high flexibility and adaptation by SMEs (Corò and Grandinetti, 1999). These geographically bound supply chain structures made firms in industrial districts able to adapt rapidly to the different requests of changing consumers and to provide the adequate degree of variety in products increasingly demanded by heterogeneous consumers. Adaptation to changing tastes of consumers was granted by the geographical contiguity to advanced customers: in industries such as fashion, furniture, jewelry, the Italian market represented for a long time the most advanced frontier of innovation in tastes and in elegance. Italian consumers, thus, anticipated the needs and demands of consumers around the world. Capitalizing on consumers' knowledge was relatively easy, given the "captive" nature of consumers themselves: they were geographically close to local firms and shared widespread and consolidated product cultures (Becattini, 1991).

The situation has dramatically changed during the last years, and globalization brought the emergence of several hotbeds for innovation in consumption practices and trends. Even though classical Made in Italy products are appreciated by a large number of consumers worldwide, new trends and tastes in fashion, furniture and other low-tech industries are being created by consumers and companies in several areas. Metropolitan areas in the USA are responsible for many of the most visible and successful casual and informal trends in clothing—for example, streetwear—China and the Far East in general are the sources of innovative trends and styles in both clothing and furniture. In other words, innovative consumers and hotbeds for innovation in tastes and cultures of consumption are multiplying and companies willing to compete in markets for fashion and home product items are called to interact with these consumers and with their communities to gather innovative ideas and knowledge, intercept new trends, create a strong association between their brands and the prevailing identities, lifestyles and cultures.

Italian small and medium companies are challenged by this strategic change: they need to re-define their approach to consumer communities and sophisticated customers, which are not geographically close nor share their same cultures, languages and social backgrounds. While the absorption of consumer knowledge was traditionally eminently tacit for Italian SMEs, today it needs to rely on explicit practices and on a strategic intent. This new approach means developing strong customer intimacy through innovative channels, the promotion of experience and interactive communication, which require a transformation in the firm's organizational competences to compete. An important instrument to develop and sustain relationships with sophisticated consumers is represented by the Internet, intended as a set of technologies, which allows companies to enter in constant and continuing relationships with consumer communities on a global scale.

On the one hand, the main leading firms' strategies carried out within industrial districts consider the introduction of complex innovation as an important element of a firm's distinctiveness (Belussi *et al.*, 2003). On the other hand, however, a completely different approach has to be developed when promoting innovation based on the involvement of customers and user communities, through social dynamics and shared sense-making. Following the Italian experience in the field of design and craft production, SMEs are re-defining their capabilities in managing manufacturing processes and enhancing their relational capabilities in the marketing domain. Information and communication technology plays a crucial role in the activation and consolidation of relationships with geographically dispersed consumers (Micelli, 2000).

In the following paragraphs, we will analyze two case studies of Italian SMEs which have started to re-define their approach to markets through stable relationships with consumer communities, both online and offline, in order to identify the strategic contribution of communities to competitiveness and innovation in Made in Italy. The two case studies were selected according to the following criteria. Both the companies can be defined as typical Italian SMEs for size (less than 250 employees), industrial specialization (fashion is a key specialization of Made in Italy) and localization in or close to industrial districts (sport system district and clothing district). A second motivation for the selection is their positive economic performances. These companies have obtained positive economic performances while the average performance of competing companies located in the same areas and operating in the same industry has been at least stable if not decreasing in the last 4–5 years as the result of increased global competition.

5.2. 55DSL: Consumer Communities and Urban Cultures in the Fashion Industry

55DSL is a company located in the area of Vicenza (Veneto region, northeast of Italy), home for many companies specialized in the production of clothing and fashion items. The company belongs to Diesel Group, a leading brand in the fashion industry. 55DSL became financially autonomous in 2005 and since then its turnover has grown from 1 million to more than 30 million euros worldwide. 55DSL was started as an experimental line oriented towards the street culture and the emerging fashion trends among young consumers influenced by the culture of skateboarding and extreme sports, later addressing also other cultures of consumption, in particular those of gamers and several music-related subcultures.

The styles and design of 55DSL products are directly influenced by the fashion trends going on in the action sports communities, and the company has sponsored many professional riders such as Tony Hawk, Joel Tudor and Steve Berra. The original inspiration coming from action sports has been then mixed with prevailing trends in urban and street culture—being the two very close, especially in the case of skateboarding clothes and urban clothes. The up-to-date design and styles associated with the company's products and brands is made possible by the continuous activity of trend scouting in several major cities which are the cradles of the most innovative trends in urban cultures around the world and that represent the focal points from which new styles and trends are diffused on international markets.

The design approach of the company recognizes the importance of the "Do it yourself" attitude prevailing today in the younger segments of fashion markets. Sophisticated consumers in the younger segments of the fashion market—aged 15–25—act like *bricoleurs* and integrate different clothing items and different styles in original and very personal ways of dressing and combining clothes. Maintaining a close relationship with young urban consumers is not fundamental only in marketing trends but increasingly useful also in order to adopt novel aesthetic and stylistic solutions into product innovation. Those solutions are developed by consumers which adapt and re-interpret existing styles and clothing lines, colors and designs as a means for self-expression and personalization.

What is generally defined as urban culture is the result of the influence of different subcultures and elements of popular culture: action sports and their aesthetics, urban fashion trends, videogames, musical subcultures. All these elements are continuously integrated in original and continuously evolving sets of identities, values and attitudes by individuals—consumers. The result of this mix is a valuable starting point for companies specialized in the production of fashion items and accessories. Interacting with these cultures is fundamental both for the strong association of a brand to prominent cultures in younger generations and for the identification of ongoing trends and potential evolutions.

55DSL in particular focused on the importance of building a bridge with communities of gamers and with communities of young music fans.

55DSL approach to online communities. Although in its early stages, the company has started to develop a strategy based on the commitment and the interaction with online and offline consumer communities since, as its management states, this could lead to capitalization on useful sources of new ideas and styles. The company website has a blog within which subscribed users can contribute with original content, ideas, suggestions and multimedia content. In the company's own words (55DSL website, see Figure 1), the website is a "user-driven non-hierarchical social bookmarking" tool through which users decide the contents that will appear on the company home page and through which they will vote and comment on the most interesting contributions. All the messages and comments of the company are categorized by users in several areas, such as arts, fashion, culture, web picks and the like.

Every week users post more than 50 messages and discussion topics and contribute with pictures and multimedia items. The idea of the whole project is, on the one hand, to make the website a primary source of useful information and interactions for young urban consumers and, on the other hand, of the dynamics and evolution undergoing in the cultures and subcultures targeted by the company.

In 2006 the company developed a partnership with EA Sports, one of the major videogame publishers in the world, in order to sponsor with its clothes the major characters of the game Need For Speed Carbon. 55DSL's website had a dedicated section to this partnership within which, besides presenting marketing material for both the game and the clothing lines, registered users can take part in online competitions on games developed purposely for the community. Every week a contest was organized and each winner was

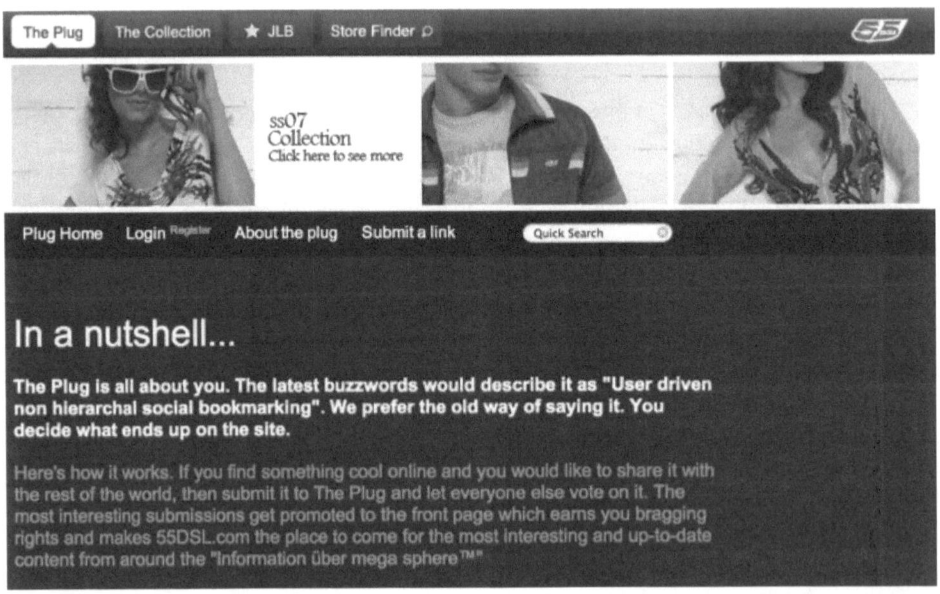

Figure 1. 55DSL website. *Source:* http://www.55dsl.com/?page=about

attributed a prize—55DSL clothes specially designed in limited editions for the Need For Speed Carbon line. Winners of the weekly competitions participated in a draw competition that attributes other prizes, always from the same limited line. Creative consumers were called to design their outfits for the characters of the game, through a sort of design context that could bring innovative solutions and suggestions related to emerging preferences and styles among the company's consumers.

The philosophy of being a brand close to the most vivid trends in the most "fashionable" cities in the world gave life also to another online initiative, called Junior Lucky Bastards. Every year the company selects two young customers and sends them around the world in cities like Paris, Tokyo, Los Angeles and the like for 55 days. The two young reporters had to take pictures, movies and music from every place and maintained a blog and a podcast. Customers in the community, through blog comments and email, suggested places to visit and things to do, and hot spots in the cities the two were visiting. As for the previous activities, this initiative had two objectives: creating a branded experience and gathering impressions and first-hand rich accounts of newer styles and trends upon which consumers on the website can comment and contribute.

Music is another important driver of the relationship between the company and the most advanced communities of consumers. 55DSL organizes several sponsored parties and musical events which are anticipated and promoted online through its website and sponsors also the production of music compilations and records by famous Italian artists which are very well known in the street and urban-oriented communities. The flagship store of the company in London organizes weekly events and musical gatherings, which are held by famous deejays and broadcasted through the Internet and through podcasts.

The strategy has proved to be potentially successful in terms of insights and ideas for new collections and lines coming from the creativity of consumers although it is not yet clear how sustainable this strategy can prove in the long run. In fact, there is no clear indication that the users interacting online with the company are actual customers of 55DSL products. The risks associated with this uncertainty are clear: insights, suggestions and ideas emerging from these interactions could be misleading in the process of development of new collections, lines and designs.

5.3. Communities of Consumers and Innovation in Sport System: The Case of Alpinestars

Alpinestars is a world-renowned brand in the motorbike boot and equipment industry, located in the sport system district of Montebelluna (Veneto region). The company started as a small manufacturer of ski boots in 1964 and shifted its focus to motorcycling boots and equipment—gloves, suits, technical protection and safety items—during the 1970s, following the relevant growth in popularity and market size of motor sports. It is actually among the leading brands in professional circuits such as Formula 1, MotoGP, Nascar, and has also developed casual clothing lines strongly inspired by the world of motor sports and street culture.

The company experienced strong growth during the last 10 years. In particular, between 1997 and 2002 the company more than doubled its revenues, passing from 40 million euros to 80, and between 2002 and 2005 it kept growing, both in terms of sales (100 million in 2004, 120 million in 2005) and in the number of employees, which account to more than 200 globally. Started as a traditional small company located in a specialized industrial district, it grew to become a major player with a strong international presence—more than half of the total turnover is generated on international markets.

Alpinestars has answered to a market which is particularly demanding under the functional and ergonomic standpoint and at the same time has succeeded in re-defining the nature of its products differentiating its lines of boots and equipment through an increasing aesthetic and "fashion" content—forms, colors, shapes and styles. Motor sports have in fact become strongly connected to cultures of consumption, which are strong drivers of identity building and sense of belonging by practitioners and enthusiasts. They demand an increased expressive content of the products they use in addition to performance. Alpinestars—as well as other Italian companies in the industry such as Dainese and Ducati in motorbike production—have been able to sense these dynamics in their markets and developed strong brands, innovative product concepts and recognizable designs and aesthetics (Verona and Prandelli, 2006).

As far as the functional characteristics of its products are concerned, Alpinestars maintains strong connections with professional pilots and lead users in motor sports—endorsed pilots in major circuits—as a source for novel ideas and novel solutions for the improvement of its products. Not only professional riders but also biking enthusiasts willing to share their competences are important actors in Alpinestars' innovation processes.

Product innovation in such an industry depends heavily on the ability to deliver subtle but extremely effective solutions in two areas: safety and comfort. In order to develop such solutions, contact with the most advanced and sophisticated users is required. Sponsoring champions has a double utility in this case: creating brand awareness and cooperating with the most demanding customers. Within the process of new product development, designers and engineers work closely, in multidisciplinary teams, in order to guarantee the perfect fit between the functional and ergonomic concepts and proposals coming from designers and the technical solutions elaborated by engineers.

While professional riders represent a fundamental source of information for new solutions development, their requests and demands are not necessarily connected to the requests of the general market. The company perceived the need to interact also with bikers and riders whose main concerns are not only related to extreme performances such as MotoGP races but to several other factors such as stylistic and design improvements and developments.

In order to tap from a valuable source of ideas and suggestions regarding the biking culture, the company moved its R&D and design facilities to Torrance (Los Angeles, California), in order to relate to sophisticated, although not professional, bikers. Torrance is the homeland for freestyle biking and motocross, and being close to local communities of riders provides a second fundamental point of observation in order to develop further solutions, observing riders in their practices and dialoguing with them. Besides the functional improvements deriving from such a relationship, the company benefited from the immersion and participation in the overall biking culture which allowed it to monitor the specific trends undergoing in biking and similar communities—extreme sports enthusiasts and skaters—and develop innovative stylistic and aesthetic solutions. These sports in fact have a strong lifestyle connotation: often their practitioners are characterized by specific ways of dressing and behaving, and by specific cultures and values.

Being immersed in the physical location where the most innovative and sophisticated consumers are allows the company to catch the latest aesthetic trends. The company's designers entertain constant relationships with bikers in the area and participate in motorcycling events as a means to take part in the shared experience space (Prahalad and Ramaswamy, 2003) within which users' experimentations with styles, fashions and designs are captured by the company and capitalized in the new product development pipeline. Alpinestars' strong connection to the most sophisticated user communities in the world helped its brand to be legitimized among those communities and become one of the favorites.

A second instrument used by the company to strengthen its relationship with user communities is the Internet. Alpinestars' website is conceived as a hub for information related to the world of motorcycling (Figure 2). Several pilots are interviewed periodically by the company which then inserts audio, video and text contributions to specified areas in the website. Video shooting of races and performances are continuously provided to the visitors of the website and Randy Mamola, a famous biker now turned TV commentator for the MotoGP, and Peter Wilson, another famous TV commentator of MotoGP races, have their columns in a specific area of the website and interact with consumers. Podcasts are regularly produced and distributed after every major motorcycling and motor sport event.

The company has developed a proactive strategy towards communities through its physical close contact with the California racing community and has adopted a different approach on the Internet. Though the website is not community oriented, since it does not offer specific areas dedicated to consumer interaction, the company has started to monitor major motorcycling blogs and websites in order to draw useful suggestions related to the wider cultural and aesthetic trends being developed in motor sports enthusiast communities around the world. Blogs and motorbiking websites, personal photographic albums and multimedia contributions of bikers documenting their experiences, the events they take part in and their consumption contexts are often important for the company in order to adjust its

Figure 2. Alpinestars website. *Source:* http://alpinestarsinc.com/moto/mamola

products' styles according to the trends emerging in consumer communities. Moreover, those insights are important to better understand the evolution of consumption cultures and consequently adapt Alpinestars' branding strategies and communication activities to maintain its legitimacy and credibility among riders and enthusiasts.

While in the case of 55DSL the company has opted for the construction of a branded space for its community—which is based on its websites and directly controlled by the firm—Alpinestars adopted a more open approach. Instead of hosting a specific community on its website, the company monitors the wider community of bikers and sports enthusiasts which is characterized by a strong identity and by a substantial loyalty towards several brands and products, as well as towards several interaction spaces—virtual and offline.

6. Conclusions and Further Research Perspectives

Our analysis on the role of consumer communities for Italian companies operating in low-tech industries unfolds promising research perspectives for further studies on the contributions of user communities in innovation processes. In particular, the consideration of user contributions to the intangible dimension of products suggests interesting integrations between theories in innovation studies and consumer studies as well as marketing in order to obtain a richer theoretical framework for the interpretation of the strategic role of consumer communities in innovation in markets for consumer goods.

The two case studies highlight insights companies can gain in terms of new ideas and feedbacks for product innovation and value creation, based on an explicit strategy of relationships with user communities. While the focus on manufacturing activities push many Italian SMEs to focus on industrial processes and, often, on functional characteristics of products, the attention to customers in their contexts of use and the understanding of social dynamics could support strategies of differentiation through the interplay of functional and intangible dimensions. In such a perspective, firms promote an intensive and interactive marketing approach, where the design of customer experiences as well as the messages and contexts proposed in their communication are built on the user's perspective, with their languages, values and sense-making results.

The role of ICT appears crucial to set up user–firm innovation circuits. Italian SMEs only recently started to recognize the potential of web technologies for the creation of rich relationships with customers. This means not only focusing on the firm's website, but also surfing the net and discovering customers' online environments, where valuable dynamics takes place.

As opposed to many other Italian SMEs specializing in medium and low-tech industries, 55DSL and Alpinestars have invested in a marketing strategy that explicitly puts lead users and communities at the core of their design and innovation processes. In addition, 55DSL has promoted an unconventional use of ICT in order to create and exploit the web as a customer management platform for experience sharing and brand interaction. Alpinestars refers to network technologies to share online experiences rooted in the physical context with customers, while monitoring existing independent communities for additional innovation inputs. Both the firms are SMEs where most of the manufacturing activities are outsourced to focus on value added processes such as product development and marketing, internally managed.

Our empirical analysis is limited to two explorative case studies, from a managerial perspective it offers preliminary issues worth addressing with further studies. First, for a firm to rely on user communities requires structured and focused marketing strategies, where key elements are brands, communication and interaction. Second, firms have to develop novel competences and professional profiles, not exclusively product-oriented, but related also to marketing, technology and sociology. Third, firms have to upgrade their technology strategies, through a movement towards more mature models of ICT adoption and in particular towards the use of social and relational technologies (e.g. web 2.0 tools, see O'Reilly, 2005; Tapscott and Williams, 2006).

SMEs could benefit from these strategies, because of their vocational collaborative approach to markets (as in the case of the Made in Italy district model) as well as the economic advantages the communities and user-generated content approach can offer, compared to structured R&D-based innovation. However, small size may become a limit in terms of competences and resource available internally—specifically in the domain of marketing or ICT—and require new forms of division of labor at the local and international level (networking as a solution to exploit external specialization).

A limitation of our study is that it is still preliminary and further analyses have to be carried out within the selected firms to explain how internal processes are changing. Moreover, an in-depth analysis should be promoted also at the community level, to understand the users' dynamics both in the virtual environment and offline. In addition, the set of incentives the firms have to develop should be further analyzed in order to obtain user's involvement into non-functional innovation processes. Future research should also focus on identifying measures able to evaluate how communities' impacts on innovation affect firm's performance. As many studies on communication and the role of intangible assets highlight, there are no systematic tools and procedures to quantify the returns of community management and their inputs on firm's economic results.

Acknowledgements

Marco Bettiol, Maria Chiarvesio, Stefano Micelli, the anonymous referees and the participants at the European Academy of Management conference held in Paris, 16–19 May 2007 whose comments have helped the improvement of this work are thanked. The shortcomings of the paper are the authors' responsibility alone.

References

Aaker, D. A. and Joachimsthaler, E. (2000) *Brand Leadership* (New York: Free Press).

Ahuvia, A. C. (2005) Beyond the extended self: loved objects and consumers' identity narratives, *Journal of Consumer Research*, 32, pp. 171–184.

Bagozzi, R. P. and Dholakia, U. M. (2006) Antecedents and purchase consequences of customer participation in small group brand communities, *International Journal of Research in Marketing*, 23, pp. 45–61.

Becattini, G. (1991) Italian industrial districts: problems and perspectives, *International Studies of Management & Organisation*, 21, pp. 83–90.

Becattini, G. and Rullani, E. (1996) Local systems and global connections: the role of knowledge, in: F. Cossentino, F. Pyke & W. Segenberger (Eds), *Local and Regional Response to Global Pressure: The Case of Italy and its Industrial Districts* (Geneva: International Institute for Labor Studies).

Belk, R. W. (1988) Possessions and the extended self, *Journal of Consumer Research*, 15, pp. 139–168.

Belussi, F., Gottardi, G. and Rullani, E. (2003) *The Technological Evolution of Industrial Districts* (London: Kluwer).

Bettiol, M. and Micelli, S. (Eds) (2005) *Design e creatività nel made in Italy. Proposte per i distretti industriali* (Milano: Bruno Mondadori).

Chesbrough, H. W. (2003) *Open Innovation* (Cambridge, MA: Harvard Business School Press).

Chiarvesio, M., Di Maria, E. and Micelli, S. (2004) From local networks of SMEs to virtual districts? Evidence from recent trends in Italy, *Research Policy*, 33, pp. 1509–1528.

Corbellini, E. and Saviolo, S. (2004) *La scommessa del Made in Italy e il futuro della moda italiana* (Milano: Etas).

Corò, G. and Grandinetti, R. (1999) Evolutionary patterns of Italian industrial districts, *Human Systems Management*, 18, pp. 117–129.

Cova, B. (1997) Community and consumption: towards a definition of the "linking value" of products or services, *European Journal of Marketing*, 31, pp. 297–316.

Di Maria, E. and Micelli, S. (2007) Imprese del made in Italy, competitività e innovazione, in: G. Volpato (Ed.), *Il knowledge management come strumento competitivo. Un confronto intersettoriale* (Roma: Carocci).

Economist, The (2007) Lessons from Apple. Available at http://www.economist.com/opinion/displaystory.cfm?story_id=9302662.

Franke, N. and Piller, F. (2004) Value creation by toolkits for user innovation and design: the case of the watch market, *Journal of Product Innovation Management*, 21, pp. 401–415.

Franke, N. and Shah, S. (2003) How communities support innovative activities: an exploration of assistance and sharing among end-users, *Research Policy*, 32(1), pp. 157–178.

Franke, N. and von Hippel, E. (2003) Satisfying heterogeneous user needs via innovation toolkits: the case of Apache security software, *Research Policy*, 32, pp. 157–178.

Hagel, J. and Armstrong, A. G. (1998) *Net Gain: Expanding Markets through Virtual Communities* (Boston: Harvard Business School Press).

Harhoff, D., Henkel, J. and von Hippel, E. (2003) Profiting from voluntary information spillovers: how users benefit by freely revealing their innovations, *Research Policy*, 32, pp. 1753–1769.

Hirschman, E. C. and Holbrook, M. B. (1982) Hedonic consumption: emerging concepts, methods and propositions, *Journal of Marketing*, 46, pp. 92–101.

Jenkins, H. (2006) *Fans, Bloggers and Gamers. Exploring Participatory Culture* (New York: New York University Press).

Jeppesen, L. B. and Fredriksen, L. (2006) Why firm-established user communities work for innovation? The personal attributes of innovative users in the case of computer-controlled music instruments, *Organization Science*, 17, pp. 45–64.

Jeppesen, L. B. and Molin, M. (2003) Consumers as co-developers: learning and innovation outside the firm, *Technology Analysis and Strategic Management*, 15, pp. 363–383.

Kotro, T. and Pantzar, M. (2002) Product development and changing cultural landscapes: is our future in snowboarding?, *Design Issues*, 18, pp. 30–45.

Kozinets, R. (1999) E-tribalized marketing? The strategic implications of virtual communities of consumption, *European Management Journal*, 17, pp. 252–264.

Krippendorf, K. and Butter, R. (1984) Product semantics: exploring the symbolic qualities of form in innovation, *The Journal of the Industrial Designers Society of America*, 3, pp. 4–9.

Lazerson, M. and Lorenzoni, G. (1999) The firms that feed industrial districts: a return to the Italian source, *Industrial and Corporate Change*, 8, pp. 235–266.

Leonard, D. and Rayport, J. F. (2000) Spark innovation through empathic design, *Harvard Business Review*, 6, pp. 103–115.

Lush, R. F., Vargo, S. L. and O'Brien, M. (2007) Competing through service: insights from service-dominant logic, *Journal of Retailing*, 83(1), pp. 5–18.

Maffesoli, M. (1992) *La trasfiguration du politique: la tribalization du monde* (Paris: Grasset).

Micelli, S. (2000) *Imprese, reti e comunità virtuali* (Milano: Etas).

Morling, M. S. and Strannegard, L. (2004) The silence of the brands, *European Journal of Marketing*, 38, pp. 224–238.

Morrison, P. D., Midgley, D. F. and Roberts, J. H. (2004) The nature of lead users and measurement of leading edge status, *Research Policy*, 33, pp. 351–362.

Muniz, A. M. and O'Guinn, T. (2001) Brand community, *Journal of Consumer Research*, 27, pp. 412–432.

Nambisan, S. (2002) Designing virtual customer environments for new product development: towards a theory, *Academy of Management Review*, 27, pp. 392–413.

Normann, R. (2001) *Reframing Business: When the Map Changes the Landscape* (Chichester: Wiley).

Nussbaum, B., Berner, R. and Brady, D. (2005) Get creative!, *Business Week*, Issue 3945, pp. 60–68.

O'Reilly, T. (2005) What is web 2.0? Design patterns and business models for the next generation of software, *O'Reilly Net*. Available at http://www.oreillynet.com.

Pine, B. J. and Gilmore, J. (1999) *The Experience Economy* (Boston: Harvard Business School Press).

Pine, B. J. and Gilmore, J. (2007) *Authenticity. What Consumers Really Want* (Boston: Harvard Business School Press).

Porter, M. E. (1985) *The Competitive Advantage: Creating and Sustaining Superior Performance* (New York: Free Press).

Postrel, V. (2004) *The Essence of Style* (New York: HarperCollins).

Prahalad, C. K. and Ramaswamy, V. (2003) The new frontier of experience innovation, *MIT Sloan Management Review*, 44, pp. 12–18.

Pyke, F., Becattini, G. and Segenberger, W. (Eds) (1990) *Industrial Districts and Inter-firm Cooperation in Italy* (Geneva: International Institute for Labour Studies).

Ravasi, D. and Lojacono, G. (2005) Managing design and designers for strategic renewal, *Long Range Planning*, 38, pp. 51–77.

Rémy, E. and Kopel, S. (2002) Social linking and human resource management in the service sector, *The Service Industries Journal*, 22(1), pp. 35–56.

Rheingold, H. (2002) *Smart Mobs* (Cambridge: Perseus).

Roberts, K. (2004) *Lovemarks: The Future Beyond Brands* (New York: PowerHouse Books).

Rosenberg, N. (1982) *Inside the Black Box: Technology and Economics* (Cambridge: Cambridge University Press).

Rullani, E. (2004) *La fabbrica dell'immateriale* (Roma: Carocci).

Sawhney, M. and Prandelli, E. (2000) Communities of creation: managing distributed innovation in turbulent markets, *California Management Review*, 42, pp. 24–54.

Schmitt, B. and Simonson, A. (1997) *Marketing Aesthetics. The Strategic Management of Brands, Identity and Management* (New York: Free Press).

Schouten, J. W. and McAlexander, J. H. (1995) Subcultures of consumption: an ethnography of new bikers, *Journal of Consumer Research*, 22, pp. 43–61.

Siggelkow, N. (2007) Persuasion with case studies, *Academy of Management Journal*, 50, pp. 20–24.

Silverstein, J. M. and Fiske, N. (2003) *Trading Up. The New American Luxury* (New York: Portfolio).

Sproull, L. and Kiesler, S. (1991) *Connections. New Ways of Working in the Networked Organization* (Cambridge, MA: MIT Press).

Tapscott, D. and Williams, A. D. (2006) *Wikinomics. How Mass Collaboration Changes Everything* (New York: Portfolio).

Verganti, R. (2006) Innovation through design, *Harvard Business Review*, 84, pp. 114–122.

Verona, G. and Prandelli, E. (2006) *Collaborative innovation. Marketing e organizzazione per i nuovi prodotti* (Roma: Carocci).

Veryzer, R. V. (2005) The role of marketing and industrial design in discontinuous new product development, *Journal of Product Innovation Management*, 22, pp. 22–41.

von Hippel, E. (1988) *The Sources of Innovation* (New York: Oxford University Press).

von Hippel, E. (1994) Sticky information and the locus of problem solving: implications for innovation, *Management Science*, 40, pp. 429–439.

von Hippel, E. (2001) Innovation by user communities: learning from open source software, *MIT Sloan Management Review*, 42, pp. 82–86.

von Hippel, E. (2005) *Democratizing Innovation* (Boston: MIT Press).

von Hippel, E. and von Krogh, G. (2003) Open source software and the private-collective model of innovation: issues for organization science, *Organization Science*, 14, pp. 208–223.

Yin, R. K. (1994) *Case Study Research: Design and Methods* (Thousand Oaks, CA: Sage).

Leveraging Lead User Knowledge in Software Development—The Case of Weblog Technology

STEPHAN KAISER & GORDON MÜLLER-SEITZ

ABSTRACT Firms increasingly rely upon information technology (IT) to manage organizational knowledge, though this does not inevitably result in increased knowledge sharing. In contrast, we know that in the case of non-commercial open software development, IT plays a central role in knowledge sharing between software developers. This paper acknowledges the impact of a blogosphere—a system of connected weblogs (i.e. personalized and informal publications on the Internet in reverse chronological order)—on the motivation of lead users to develop commercial software together with the employees of a large for-profit organization. On the basis of multi-method data, collected over a 28-month period, our results indicate parallels to the field of open source software: we argue that weblog technology and its features evoke intrinsic and extrinsic motivation to engage in knowledge sharing within a commercial software development project.

Each man's actions are influenced by his special opportunities and resources, as well as by his temperament and his associations. (Marshall, 1920: 355f.)

Introduction

Today, the importance of information technology (IT) for the success of most organizations is widely acknowledged. Likewise, knowledge is deemed a key strategic asset for organizations (Nonaka, 1994; Turvani, 2001). Bearing these considerations in mind, it is not surprising that organizations endeavor to combine IT and knowledge management activities in order to derive potential benefits (Davenport and Prusak, 1998). However, earlier studies have shown (Davenport, 1997; Raub and von Wittich, 2004) that oftentimes investments in IT do not facilitate knowledge sharing and organizational learning within the firm, since employees are not motivated to share their knowledge (Constant *et al.*, 1994; Ciborra and Patriotta, 1998). Therefore, one can argue that the integration of external knowledge and expertise (e.g. Leiponen, 2005; Lane and Probert, 2007) via IT might be even more difficult if there are no market-based incentives for external people to contribute their individual knowledge to the firm. In contrast, we know that in the case of commons-based peer production (Benkler, 2002) and in the case of non-commercial open source software (OSS) development (Raymond, 2001; Lakhani and von Hippel, 2003), highly qualified individuals engage in knowledge activities beyond hierarchical coordination. It is argued that contributors to OSS are foremost intrinsically and also extrinsically motivated (Lerner and Tirole, 2002) to share their knowledge by means of improving software codes. Thus, OSS might be viewed as an exemplary novel way of organizing knowledge and capabilities (Langlois and Robertson 1995; Loasby, 1998).

Against this background, we aim at investigating the impact that a particular IT exerts on the motivation of employees and external experts that dispose of lead user features (von Hippel, 1989) to share and contribute knowledge during the development of a commercial software. Thereby, we attempt to observe the integration of decentralized and hierarchical coordination of knowledge activities. We analyze the case of weblogs—that is, personalized and informal publications on the Internet in reverse chronological order—and the motivation of webloggers to share their knowledge and ideas. Our arguments are based upon the distinctive features of weblogs that transform this technology into a tool for organizing distributed knowledge and stimulating ideas from outside the boundaries of the firm (Dahlander, 2005; Chesbrough *et al.*, 2006). We illustrate that a networked system of individual weblogs—subsequently referred to as blogosphere—improves the access to external knowledge and to innovative ideas from an organizational stance. Thereby, we take into account arguments of Loasby (1998, 2001) and suggest that the organization of knowledge is strongly connected to the organization of incentive structures and motivation. On the basis of this consideration, the objective of this paper is to address and combine the following research questions:

Research question 1: What are the intrinsic stimuli that motivate lead users to support the software development of a for-profit organization by utilizing weblog technology?
Research question 2: What are the extrinsic stimuli that motivate lead users to support the software development of a for-profit organization by utilizing weblog technology?

As a result, we show that even the motivation of external experts to contribute knowledge to a for-profit software development follows similar arguments as is the case with the engagement in OSS development. The features of weblog technology have considerable impact on the exchange of knowledge across organizational boundaries because the features support the emergence of appropriate motivational stimuli.

The procedure of this paper is as follows: first, we introduce insights from the field of open source software in connection with social psychological accounts as the theoretical foundation of our argument. Second, the general features of weblogs as novel communication tools are presented and contrasted with traditional online communication devices. In addition, a synopsis of our longitudinal field research is delineated. The consolidated findings shed a new light on the way benefits can be derived from innovation processes that tap into sources from outside the organization. Thereby, we complement theoretical considerations with insights from our empirical findings. In particular, we elucidate the mechanisms that culminate in a lead user leverage, a term that alludes to the fact that lead users were comparatively more engaged than the employees of the respective corporation. Finally, we present the contributions of our findings, limitations of this research and future research options.

Sociopsychological Accounts within Open Source Software Research

One particular area where intriguing motivational stimuli are aroused concerns the development of OSS, a different approach to develop computer software in contrast to conventional proprietary solutions (Raymond, 2001; Lakhani and von Hippel, 2003). Although differences exist between the various OSS initiatives, the term "open" indicates that anyone is permitted to study, change, improve and distribute the unmodified or modified software. Thereby, "open source" implies that the source code of the software is freely available, a stark contrast to the conventional, proprietary solutions. The overall goal of these communities is the development of a commonly shared software, that is, a knowledge-intensive good that is jointly produced. Some of the most prominent examples are Apache, Linux and Mozilla. For our study purposes, it is particularly worth noting that many top-notch software developers make their knowledge openly accessible and, as a rule, without any remuneration for these communities. Since the contributors to OSS projects are not formal members of an organization, a novel explanation for the motivation of these participants appears to be essential (Lakhani and von Hippel, 2003).

Referring to the experimental studies of Deci (Deci, 1975; Deci and Ryan, 1985) in the field of social psychology, one can borrow the dualistic concept of intrinsic (i.e. activities and behaviors that people naturally engage in for their own sake) and extrinsic (i.e. where direct compensation for the work or actions a person undertakes is expected) motivation (cf. Lerner and Tirole, 2002; Dahlander and McKelvey, 2005; Roberts *et al.*, 2006 for related conceptions and discussions). Both intrinsic and extrinsic motivational stimuli are a prerequisite for successful knowledge sharing between the diverse partners (for a related discussion: Osterloh and Frey, 2000; Osterloh *et al.*, 2002; Sorenson and Singh, 2007).[1]

Regarding intrinsic motivation, the contributors to OSS projects often indicate that they simply enjoy creating an improved part of the source code (Levy, 1984; Torvalds and Diamond, 2002; Lakhani and Wolf, 2005). Thereby, an innate desire seems to be of utmost importance, as this might lead to a greater possibility of goal attainment (Sheldon and Elliot, 1998). In this connection, other authors have argued that the enjoyment stems from the

[1] The debate about intrinsic and extrinsic motivation has been refined over the years since Deci published his seminal work in 1975. In particular, researchers discuss potential crowding-out effects of extrinsic motivational stimuli upon intrinsic motivation, for example, Frey (1997) and Frey and Oberholzer-Gee (1997) from a behavioral economics perspective or Amabile (1997) from a managerial stance (for recent reviews, cf. also Deci *et al.*, 1999; Benabou and Tirole, 2003). In this paper, we explore the existence of both intrinsic and extrinsic motivation of lead users without discussing potential crowding-out effects due to explorative focus upon the stimuli that lead to a voluntary engagement of external experts.

intellectual challenge (Hars and Ou, 2002; Lakhani and von Hippel, 2003). As a consequence, it is also conjectured that these project members will spend a considerable amount of time and effort in order to improve the source code at stake (Lakhani and von Hippel, 2003).

Moreover, an intrinsic stimulus can be the ideology surrounding OSS projects (Dahlander and McKelvey, 2005), often manifested in a libertarian way (O'Mahony, 2003; Mikkonen *et al.*, 2007). Although parallels can be drawn to the way academics converse (Bergquist *et al.*, 2001), this approach is comparatively novel with regards to the organizational arena, where the notorious hierarchical and bureaucratic elements still prevail or dominate (Hannemyr, 1999; Moon and Sproull, 2000). The central motive is to antagonize capitalism that developers deem to be epitomized by huge software corporations, such as Microsoft or Intel (Himanen, 2001; Moody, 2002). Epigones of the OSS movement—such as Stallman—therefore argue that software developers ought to be able to develop software freely (Stallman, 2001; Lin, 2003).

Furthermore, OSS researchers argue that the benefits of improving one's own code serve one's own intrinsic needs (Lerner and Tirole, 2002; Franke and von Hippel, 2003; Weber, 2004). In a similar vein, other studies have related this to the observation that it signifies that the person can exert an innate impact upon its environment, improving one's own self-efficacy (Kollock, 1999; cf. also Bandura, 2000). This implies in the context of OSS projects that developers assume to be able to contribute to collective goals by means of improving the software code.

Hooking up to the previous intrinsic stimuli, reciprocal social exchange manifested in altruistic and prosocial behavior—an intrinsically motivated person seeks to augment the welfare of other people—can lead to contributions (Kollock, 1999; McLure Wasko and Faraj, 2000; von Krogh *et al.*, 2003; Zeitlyn, 2003). The results from other studies also suggest that such behavior is linked to a sense of belonging to a community (Lakhani and Wolf, 2005). Therefore, the developers feel obliged to interact with the community by contributing their knowledge (Simon *et al.*, 1998; Hertel *et al.*, 2003; Bagozzi and Dholakia, 2006; for a critical perspective: Weber, 2004).

In addition to intrinsic motivational stimuli, present or future personal rewards that stem from external influences represent extrinsic motivational stimuli. OSS researchers identified that social recognition among peers serves as an extrinsic stimulus (Lerner and Tirole, 2001). For example, while constantly updating the source code with valuable additions, a project member might enhance his/her reputation in the respective community (Lerner and Tirole, 2002), which can be labeled as a novel form of self-marketing or status and competence signaling (Lakhani and von Hippel, 2003; Jeppesen and Frederiksen, 2006). This is due to the fact that one's own knowledge becomes visible, while the evolution of thoughts remains accessible within the discussion threads by the posting of information regarding the improvement of the code (cf. in a related vein: Holmström, 1999). Regarding such status signaling, it is also argued that career opportunities in future are pursued, while constant contributions might be recognized by commercial software companies as a valuable activity (Lerner and Tirole, 2002).

Support from fellow developers presents a further external stimulus (Kollock, 1999). Due to the sharing of code lines, a strong identification is likely to ensue among the developers in the form of the ideological impact sketched above, and also the developers help each other when problems occur by writing code lines, which can be interpreted from an anthropological perspective (Malinowski, 1922; Mauss, 1990) as a practice of gift giving (cf. also Bergquist and Ljungberg, 2001; Zeitlyn, 2003). The support need not be directly reciprocal, that is, one person helps another and expects to benefit from it in the future in terms of obtaining help from the same person that received help beforehand. OSS projects are rather characterized by a generalized reciprocity that is impersonal in nature (Gouldner, 1960; Raymond, 2001). This implies that the developers support each other without expecting any short-term or direct response in the form of feedback or assistance; they rather assume that someone from the community as such will assist him or her when he/she experiences difficulties in developing the software code.

Research Design

We employed a multi-method case-study design, collecting data over a period of 28 months, using multiple methods. In the following section, we first sketch the characteristics of the weblog technology. Second, we introduce our research object, the Microsoft Longhorn Blogosphere. Third, we outline the structure of the blogosphere. Fourth, we delineate our content analysis of the weblogs that focuses on motivational issues.

Weblogs and their Distinctive Characteristics

Jorn Barger coined the term "weblog" on his Robot Wisdom website in 1997. The neologism stems from the contraction of the words "web" and "log". Concerning the definition of weblogs, Dafermos (2003) proposes several constituent features. First, weblogs refer to websites that publicly display individual thoughts. The frequencies vary profoundly, although entries are added periodically in the form of a log in reverse chronological order. In most cases, a single person renders the initial content of the site. In addition, readers can write in designated sections accompanying commentaries that often entail vibrant discussions analogous to the discourses that can be observed in the form of threads in the field of open source software development (Herring *et al.*, 2005). Moreover, each post (also called entry) is time-stamped and archived so that old content remains accessible and attributable to the author. The entries are predominantly text based and usually possess a title in the form of a large header, followed by text fragments that are often accompanied by pictures, and more recently by audio or video formats.

However, the most striking feature concerns the ability to establish diverse virtual references to other weblogs by means of various functions, namely, blogrolls, permalinks and trackbacks. Blogrolls constitute a list of links to those weblogs that are recommended by the author and those that are being read frequently. These links are usually located on one side of the respective weblog and enable the reader to infer information about the preferences of the author. In contrast, single static links, called permalinks, refer to other posts or even particular sections of a post in other weblogs or websites and emanate from a particular section within a post. In this sense, permalinks represent a modification that is closely related to the blogroll feature or the "bookmark" feature in common office-software applications. In a similar vein, in- and outbound trackbacks are a third form of coalescence. Trackbacks allow webloggers to see who has referred to his or her original post by writing a reply. This feature habitually appears below a post and shows a summary of what has been voiced on the target weblog in conjunction with the URL and name of the respective weblog.

The various means of interlinking constitute a phenomenon termed blogosphere, alluding to the fact that the entire number of weblogs in the Internet represents a distinctive IT-instantiated media ecosystem (Herring *et al.*, 2005). Because of the interlinking, there are multiple intertwined weblog networks. Although the term blogosphere encompasses, according to our understanding, all existing weblogs, within this complex system, clusters of various smaller blogospheres exist. This observation seems feasible since some weblogs are densely connected, frequently around a certain topic or field of interest similar to communities of practice (Lave and Wenger, 1991; Amin and Cohendet, 2000; Ardichvili *et al.*, 2003).

Apart from this active means of referencing and interlinking, there exists another form, which we refer to as passive referencing. This alludes to the possibility of being automatically informed about new posts or comments aired on other weblogs. Although this technological format is also quite common in other contexts, for instance, on the websites of companies such as CNN (http://edition.cnn.com/) or Reuters (http://today.reuters.com/news/default.aspx), we focus upon those with regards to weblogs. Due to one's own preferences, it is possible to gather new information by subscribing to the respective weblogs using RSS (abbreviation for "Rich Site Summary" or "Really Simple Syndication"). RSS represents a file format that serves as news aggregator free of charge and provides the news almost instantaneously to individuals who have subscribed to this service. A prerequisite for this form of referencing is that the respective weblog is subscribed to a kind of meta-weblog directory. Consequently, these directories are informed via the transmission of signals (so-called pings) each time a new post has been added to the weblog that has been registered.

Following the introduction of the conceptually distinctive characteristics of weblogs and blogospheres, one ought to separate weblogs from related Internet phenomena for further clarification. As mentioned above, augmenting weblogs with virtual references is a distinctive facet of weblogs, although the line is unfortunately quite blurred. As opposed to personal web pages ("home pages"), weblogs are commonly more frequently updated and usually more personal, particularly concerning organizational settings. Furthermore, they usually establish a relationship between the individual author and the person or community that reads the content of the web page, for instance, due to repeated exchanges by means of comments on posts and subsequent responses to them in the form of posts, etc. Other comparable phenomena are bulletin board systems, online communities, newsgroups and chats. However, they usually lack the ability to be interlinked in the multiple ways weblogs are. Moreover, they differ consistently concerning the authority to add original content. While in the other four forms the content is created jointly, in the case of weblogs, the content is usually solely creator defined. This allows webloggers to identify themselves to a larger extent with their individually designed and solely operated weblogs, which also differ from the concertedly operated wiki systems or online communities. In conclusion, we assume that weblogs are not an entirely novel phenomenon. However, weblogs can be asserted to have their own position in the broader genre ecology of the Internet, combining features of diverse online technologies, particularly the ability to have bi- or multidirectional communications in various appearances (Herring *et al.*, 2005).

The Microsoft Longhorn Blogosphere as a Research Object

As a research object, we chose a prominent example of a corporate blogosphere, the documentation of the development of Microsoft's new software "Microsoft-Longhorn" (recently launched under the name "Vista"). The Microsoft Longhorn Blogosphere (MLB) was mainly chosen for two reasons: first, in contrast to other blogospheres within the corporate realm, the weblog entries of the MLB are publicly accessible (http://www.longhornblogs.com). Thus, it was possible to examine the blogosphere properly. Second, the MLB focuses on knowledge management, as webloggers are noting information about their experiences and progress at developing and using the specific software.

Regarding the websites that purposely deal with Microsoft-related contents, one can distinguish between websites that are administered by Microsoft and those that are not. The MLB belongs to those that are not officially administered by Microsoft. However, Microsoft encourages its employees to engage themselves in the diverse company-related communities that have either an informative or a software-related scope. In addition, it is worth noting that one Microsoft employee administered the blogosphere and granted accession. The Microsoft employees and the respective lead users had to enroll themselves by contacting the administrator in order to take part in the community. The information exchanged is restricted to the Longhorn software. A prior test version was assigned to the MLB members in September 2003 in the run-up to the Professional Developers Conference (PDC) in Los Angeles. With regard to the PDC, the MLB was mainly active between September and October 2003. The intention was to provide potential lead users, voluntary contributors and developers with a platform to air their views on the upcoming software. The weblogs are concertedly operated at the respective website. The various members communicate via this platform, but they are also partially interlinked via their private weblogs. Regarding the scope of the paper, we solely focus upon those weblogs that can be subsumed under the MLB.

Microsoft employees (labeled as "experts") as well as non-Microsoft members, so-called Most Valuable Professionals (MVPs), participate as webloggers in the MLB. According to Microsoft's home page, the MVP Program

> recognizes and thanks outstanding members of technical communities for their community participation and willingness to help others. The program celebrates the most active community members from around the world who provide invaluable online and offline expertise that enriches the community experience and makes a difference in technical communities featuring Microsoft products. (Microsoft, 2005)

We argue that these MVPs act as very specific and highly knowledgeable lead users.

Analyzing the Blogosphere as a Structured, Virtual Knowledge Network

To examine the communicative structure of the MLB, we analyzed the number of posts and comments of the various webloggers. In the first instance, we identified 60 registered members, 43 of them MVPs and 17 Microsoft employees, whereas 36 of them were active webloggers (25 MVPs and 11 Microsoft experts). We were able to spot them since they identified themselves by means of their weblog entries properly as Microsoft employees or MVPs. For our study purpose, only those MLB members who contributed at least once to the MLB via a permalink, post or comment were labeled as "active" webloggers. We initially gathered the data from the Microsoft Longhorn website. The various modes of communication (posts, comments, etc. within a period from September 2003 to March 2005) were collected and entered into a spreadsheet by means of diverse matrices, whereby each of the 36 active webloggers was assigned one column as well as one row. Each dyadic relationship could be either symmetrical (also termed reciprocal) when both parties communicated with one another or asymmetrical when just one person contacted the respective MLB member. As an exemplary result, Figure 1 represents the various linkages among the MLB members, whereby in this case no differences were drawn between the types of communication (e.g. posts vs. comments).

From our data, we inferred that the MLB inherits a few central actors who have power and influence over the actors of the network. This assertion is not only based upon frequent remarks by our interviewees, but also derived from the observation that a few webloggers were most active in terms of posts and comments. For example, four out of the 32 webloggers (three MVPs and one Microsoft employee) wrote 64 percent of all the comments (Koch and Schneider, 2002; Franke and von Hippel, 2003; Kuk, 2006). This can be viewed as an indication of the fact that some webloggers are more engaged in debates than others. In addition, the posts of the most active webloggers initiated lively discussions since they provoked an above-average number of comments. If we understand the MLB as a social network, these webloggers act as brokers of knowledge and innovative ideas. A

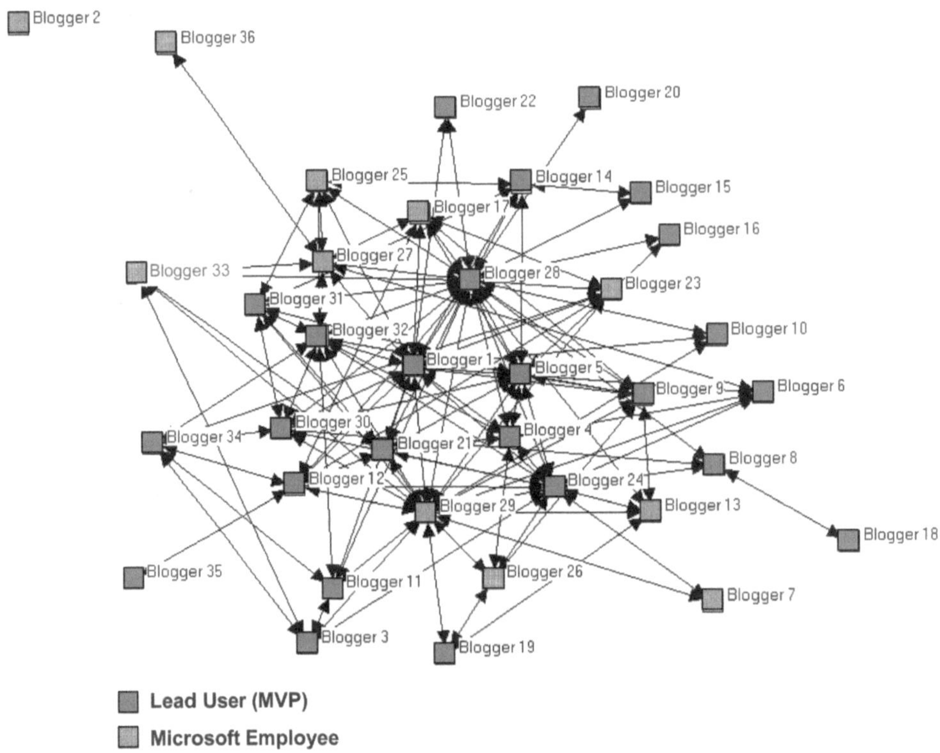

Figure 1. Illustration of the communication structure of the MLB

Microsoft representative confirmed that several of these brokers are rated to be lead users from Microsoft's point of view (May, personal communication, 22 June 2007).[2]

Analyzing the Content of Weblogs of the Blogosphere

In order to comprehend the content and motivation of weblogging, we subsequently conducted a qualitative content analysis and conducted interviews. This design can enhance the contextual understanding of a "contemporary phenomenon within its real-life context, especially when the boundaries between phenomenon and context are not clearly evident" (Yin, 2003: 13) and in situations where multiple sources of evidence are used. Parallel to the content analysis, we conducted 16 interviews. Ten of the interviewees were MLB webloggers and three were OSS developers, and we contacted the Microsoft officials thrice. In addition, we collected and analyzed 4,810 text passages (695 posts and 4,115 comments) that were posted from September 2003 to March 2005. Atlas.ti, a software solution developed for explorative research in allusion to grounded theory (Glaser and Strauss, 1967), did support our research process. The research process took place in four stages.

First, two authors read all of the 4,180 text passages and assigned open codes to themes emerging from the entries (cf. Strauss and Corbin, 1990; Miles and Huberman, 1994). The coded passages could vary considerably in length, from being only a single key term (e.g. "emoticon") to a whole post. For instance, one post dealt with a brief reflection about the development of the project, which we did not break down into separate categories at first. However, most of the time, parts of a sentence or sometimes whole sentences were chosen initially. At this stage, there was little interpretation of the information provided in the text passages.

Second, the text units were condensed in the course of our inductive analysis into 30 different categories. From these initial categories, some are still existent in the final version of this paper (e.g. "altruism/gift giving"), whereas others have been turned down due to a disagreement between the coders (e.g. "enrolment of novel participants in the MLB"). In the case of categories where no consensus was achieved, the emerging labels were modified in order to take them into account. For example, webloggers were enticed to engage in the MLB by getting in touch with and exchanging ideas with like-minded IT professionals. Thus, the category "enrolment of novel participants in the MLB" was finally collapsed into the motivational stimulus of "reciprocal social exchange". Such internal checks on the validity of the data are deemed inevitable in order to develop inductive categories (Kirk and Miller, 1986). Throughout this stage, we attempted to assign categories that took the shape of key terms. Therefore, this procedure contained our interpretation of the data, although we tried to be accurate and wherever possible, we retained key terms that stemmed from the expressions of the webloggers.

Third, by means of a selective coding procedure (Strauss and Corbin, 1990), we consolidated our categories into the five motivational stimuli presented in Table 1. In the course of this stage, we not only identified underlying patterns but also grouped them in

[2] In the course of the research process, the authors contacted Microsoft for clarification and confirmation of the findings. In this connection, a spokesperson from Microsoft Germany, May, was approached.

Table 1. Evidence from illustrative quotes for intrinsic and extrinsic stimuli

Categories of motivational stimuli	Descriptions of motivational stimuli	Quotes
Intrinsic stimuli	High degree of freedom	"LonghornBlogs.com was named the 7th best independent blog by TechWeb.com. There was one quote in particular which I was very proud of, because it was exactly what we were going for: 'Low noise, high quality, relevant independent content.' There is no way to respond to that but to smile. Thanks for making this site rock!"
	Having impact	"We are the ones who get to see our visions become a reality. Imagine not feeling like your designs were going to be trashed with compromises when it comes time to implement them. No more hacky workarounds to pull off the visual effects that you intend. Rather [...] designers get the joy of working with elegant solutions that compliment the way that your visuals were dreamed up in your favorite design tool."
	Reciprocal social exchange	"I am having the time of my life at PDC! There are so many friends that I have meet here that I seem to always meet again and again. Funny that there are like 8,000 people here, but I keep running into everyone! [...] Rockin!"
Extrinsic stimuli	Signaling competence	"My goal with my blog is to get the word out on some of these cool features and showcase cool ways to use them. I'll be posting as many common designer tips and tricks as I can. Feel free to challenge me with some cool ideas that you would like to see pulled off in Longhorn."
	Getting support	"OK Guys, tell me what you think. I'll take your feedback over the next few days to the rest of the Interscape Management Team, and we will use it to make our next moves. Thanks for your help in making LonghornBlogs.com so great. I'm looking forward to your responses."

relation to one another, that is, coded them axially (cf. Miles and Huberman, 1994). Thereby, we obtained information from prior research in the OSS arena (e.g. Lerner and Tirole, 2002), resulting in the two primary categories of intrinsic and extrinsic motivation that guide this research. Nevertheless, we remained amenable to categories that have not been the focus of prior motivation-related research enquiries with regards to the OSS phenomenon. For instance, the stimulus "high degree of freedom" has been observed to be rather seldom in previous studies though our respondents strongly confirmed this observation. Due to the longitudinal research design, we were able to reconfirm this by contacting respondents repeatedly, both the interviewees that aired this observation themselves, as well as those that we approached with this topic for asking their opinion, that is, those that have not mentioned it beforehand. As a result, we incorporated this category into our final scheme.

Fourth, in order to ensure the accuracy of the category coding, we re-examined the coding procedure for the final categories by means of a peer rating. This intercoder procedure was deemed necessary for improving the reliability of the data. While achieving high intercoder reliability with the ratings of 87 percent, we stopped the coding. Such a level of agreement is deemed appropriate for an explorative research design. By nature, this might include chance agreements (Zwick, 1988).

The validity of the data was constantly ensured by the constant comparative method (Miles and Huberman, 1994). This is necessary for qualitative research, wherein data are collected and coded inductively. Such a process has to be conducted tentatively, whereas the inductive insights need to be adjusted with information obtained from previous studies. Throughout the research process, inconsistencies arise inevitably. In order to obtain comprehensible results in terms of categories that emerge from the data, we conducted various checks (Kirk and Miller, 1986). In particular, we tried to combine archival data (i.e. post and comments) with information gathered from our interviews, and we also utilized additional data stemming from newspaper and online documents for triangulation. In addition, we repeatedly reverted to our interviewees for clarification. The respondents not only helped us to enhance our own understanding of the research context, but also confirmed or rebutted our assumptions. Here we report only those categories that were confirmed consistently by our interviewees in accordance to our interpretation of the data retrieved from the weblogs.

Discussion of Results: The Motivational Impact of Weblog Technology on Knowledge Sharing by Lead Users

We argue that Microsoft employees and lead users are motivated with regard to knowledge conversations, thereby sacrificing their resources in the form of time and thoughts. Thereby, these questions will be answered by introducing two central themes. As for the first research question, we revert to flow experiences. We propose that this intrinsic motivational stimulus emerges while knowledge in the blogosphere is being shared as webloggers read, comment and write weblog entries. With regards to the second research question, we suggest subsequently that the weblog technology provides rather extrinsic and social stimuli that foster the engagement in the blogosphere.

Flow Experiences as Intrinsic Motivation for Knowledge Sharing

According to our observations, the motivation to engage in knowledge work with regard to weblogs is well illustrated by the experience of the so-called flow states (Csikszentmihalyi, 1997), which hook up to the enjoyment often found in the accounts on OSS developers' work (Lakhani and Wolf, 2005). Flow states have not been discussed yet in an online setting with reference to multiple users, but have been discussed only with reference to single users (e.g. Ghani and Deshpande, 1994; Hoffman and Novak, 1996; Agarwal and Karahanna, 2000; Pace, 2004). Flow is a concept that can be credited to Csikszentmihalyi (1975) who coined the term while trying to find out what elicits the pleasure that people experience when they are immersed in everyday activities, creative or adventurous ones in particular (e.g. playing music or rock climbing). Flow states are self-motivating experiences that can be characterized roughly as an intense and focused concentration, whereby action and awareness are merged so that temporal awareness is distorted, an observation that has been termed autotelic (Nakamura and Csikszentmihalyi, 2002). Such flow states are primarily experienced while work is performed as opposed to leisure time (e.g. Csikszentmihalyi and LeFevre, 1989), because compared with leisure time, the nature of the work is supposed to be more challenging (Emerson, 1998).

With regards to the scope of this paper, we argue that flow can also be experienced in the course of weblogging, while a weblogger is challenged by a demanding task that is in balance with his particular skills. In what follows, we argue that three categories of activities within the MLB foster the occurrence of flow states:

- Activities allowing for a high degree of freedom in choosing one's own goals in terms of knowledge work.
- Activities enabling the weblogger to exert an impact by means of their own deeds.
- Activities facilitating a reciprocal social exchange with fellow webloggers, leading to feelings of affiliation and identity.

High degree of freedom. From our stance, the freedom to decide upon one's own involvement regarding the usage of IT is crucial for evoking people's intrinsic motivation. Triggers can take manifold forms. Inspired by discussions of Gagliardi and Strati, we claim that the aesthetic experiences of weblogging are likewise considered important (Gagliardi, 1996; Strati and de Monthoux, 2002). For instance, so-called emoticons, defined as combinations of marks and symbols in miniature sideway faces are frequently inserted into the discussion within posts or comments. In addition, capital letters are commonly perceived as expressions of emphasizing one's feelings, such as anger and joy. These forms of exhibiting oneself are viewed as a way of revealing the sender's mood (Walther and Tidwell, 1995), as the following quote illustrates:

> *I know I'm not an "old timer." I've never seen a punch card. (Although I've read Cryptonomicon)*  *I'm Generation X or Y, not Generation U or V.*

Furthermore, aesthetics are noteworthy due to the unique design features of weblogs that further elicit intrinsic motivation. For example, the user interface can be crucial since it can be designed according to the aesthetic preferences of the respective weblogger, which makes weblogs different from most of the other software solutions. This consideration is divulged by one weblogger as follows:

> *I walked away and came back to this beautiful screen: I LOVE THIS! ... Beautiful and installed!*

In addition, we propose to extend the traditional flow-related research in connection with online environments, insofar as we pay attention to the challenging nature of the task content. According to our observations, this is a relatively neglected issue in the IT arena while examinations of flow states traditionally focus on aspects of usability or functionality (Rettie, 2001). In this relationship, we observed that webloggers seem to experience flow due to the challenging nature of the tasks throughout the entire process of writing and reading, that is, accumulating or creating knowledge at the same time (Brand and Leckie, 1988; cf. also Hars and Ou, 2002; Lakhani and von Hippel, 2003). This aspect does not lead to feelings of being challenged, but users of weblogs often remark that they experience satisfaction while adding entries to their weblogs (Dafermos, 2003). Hence, similar to the field of open source software, such an innate desire is deemed essential (Sheldon and Elliot, 1998). Thus, we assume that webloggers will spend a considerable amount of time and effort, just because they derive pleasure from discursive practices such as keeping their weblogs up to date, adding entries, and reading other weblogs. In this spirit, one weblogger noted:

> *I spent a ton of time reading blogs and newsgroups and trying to help clarify things when I could. I loved talking to people about Avalon.*

Closely connected with the freedom to decide about one's own engagement within the MLB is the fact that the weblog technology's features allow for situations where freedom of speech is achieved (cf. Baldwin and Clark, 2006 and Franke and von Hippel, 2003, for the importance of the technical design in OSS projects). In view of the present context, freedom of speech is considered as the ability to air one's own voice without any hindrance. These observations have also been made with reference to the development of open source software, where one of the success factors resides in the rather anarchic nature and idealism of these projects (Hannemyr, 1999; Moglen, 1999; Moody, 2002). In the present case, the extent to which freedom of speech, and thus, the discursive making of new perspectives were enabled by the technology is extraordinarily high. Various observations support this assumption. For instance, discussions are not confined to particular topics, and there is no formal authority in place that governs the blogosphere, and each weblogger is at liberty to take part in the blogosphere in the way he/she prefers to. As for the last aspect, webloggers are able to participate in discursive practices actively, but just observation of the ongoing conversations is also accepted. The following comment from a weblogger captures this notion:

> *It's a more open and an opinionated community that is able to share their thoughts and ideas without any hindrance. Websites have become so automated and predictable, almost like zombies. I consider blogs as websites with a warm spirit within it.*

In addition, the freedom of speech is also officially manifested and at the same time secured by the mission statement of the MLB, which is another indication of the lack of a formal authority:

> *Welcome! This is NOT a Microsoft site. It is a community-based initiative to spread the word about the next version of Windows, sponsored by Interscape Technologies*

Having impact. Another determinant is one's own ability to have an impact within the MLB in terms of improving the features and design of the software while contributing to the knowledge exchange within the MLB. Stemming from the freedom of speech is the observation that webloggers can, to a certain extent, circumvent traditional hierarchies. For instance, webloggers that were extraordinarily active could become influential in providing new perspectives and changing or initiating practices (cf. also Hannemyr, 1999; Moon and Sproull, 2000). They were able to have a substantial impact on the development of the software, without necessarily being a Microsoft member or being in a dominant position within the Microsoft hierarchy, an aspect that was likewise confirmed by a Microsoft spokesperson (May, personal communication, 22 June 2007). Our analysis shows that a single weblogger achieved such a prominent and powerful position in the networked blogosphere that he/she posed as a "technical evangelist". Being able to alter the course of the software development and being able to have an impact on the practices of fellow webloggers, their motivation levels can be stimulated further, which is well voiced by one weblogger who aired that he is weblogging for

> *specific reasons, something that interests, something I have a passion for and it's also as a result of what's going on in the community. [...] I see blogging as a way for me to be able to be a driving force that can help to make a point or address a problem. Basically, blogging is my way of spreading my gospel.*

Besides, by means of the technology, webloggers were able to spread their ideas not only free of hindrance, but also very quickly. Features such as RSS lead to a very quick diffusion of information within the MLB and webloggers seemingly rely upon these technological features. As one weblogger aired:

> *We're going to be doing a really big expansion, to bring you even more Longhorn-related content. So, if you write Longhorn-related blog entries, please make sure you're categorizing your posts, and leave me a comment with your Longhorn-specific RSS feeds.*

Besides diffusion speed, the overall range of the potential audience was truly huge. Because people read each other's weblogs frequently and invest considerable amounts of time as revealed above, the ability to reach a large number of people might be a further trigger that elicits intrinsic motivation for conversation on practices, as the following post epitomizes:

> *OK Guys, tell me what you think. I'll take your feedback over the next few days to the rest of the Interscape Management Team, and we will use it to make our next moves. Thanks for your help in making LonghornBlogs.com so great. I'm looking forward to your responses*

Reciprocal social exchange. Bearing in mind that "knowing is a collective accomplishment" (Gherardi, 2003: 352), likewise, the reciprocal social communication between the webloggers can evoke flow states. We, therefore, presume the blogosphere to be a network where social participation by means of knowledge-intensive conversation regarding software development occurs (cf. for a related discussion of societal spaces: Nohara and Verdier, 2001: 217f.). In this context, one post reads as follows:

> *I spent a ton of time reading blogs and newsgroups and trying to help clarify things when I could. I loved talking to people about Avalon*

Hence, we claim that while receiving and giving feedback, the webloggers have the chance to connect and socialize themselves to a community with its own identity, which is assessed to be intrinsically rewarding (Kollock, 1999; McLure Wasko and Faraj, 2000). Various features of the weblog technology, such as comments and trackbacks, facilitate this kind of networking, which is social in nature while people get connected.

We argue that the possibilities to get visibly and permanently connected partly substitute intensive direct communication. Therefore, not only the reciprocal social exchange but also the apparent connections via technological artifacts foster the impression of belonging to a community of fellow webloggers. In this case, our analysis showed a positively connoted belief among webloggers within the MLB. Having a common objective, they seem to share collective knowledge, as well as their motivation (McArthur and Bruza, 2001). The participants of the MLB raised this issue consistently as a decisive advantage of weblogs or blogospheres in contrast to other existing online communication tools, as in the "warm spirit" quote presented above. The following post also illustrates this point:

> *This Community Rocks! Hello! First of all a huge thanks to Robert McLaws for the site, my blog, and the introduction! I really like this site and I think it's a wonderful idea, I am extremely happy to be a part of this growing community.*

Another weblogger aired regarding the warm spirit within the community:

> *What I'm excited about here is getting to shout about what I think Windows needs. So that's mostly what you'll hear from me [...] Also, it's nice to meet people like all ya'lls. It's almost like I've got ... friends ...*

Extrinsic Stimuli for Knowledge Sharing

Beyond the experience of flow, we can report evidence of extrinsic motivation for sharing knowledge via weblogs, which serves to answer research question 2. In this regard, the interviewees and the content that was aired in the weblogs suggest that weblogging advances knowledge sharing and communication in a way similar to the processes that were analyzed for the open source software community (Lerner and Tirole, 2002; Benabou and Tirole, 2003):

- First, we find arguments that webloggers talk about their knowledge in order to signal competence, which in turn makes them gain direct recognition, as well as long-term reputation.
- Second, the analysis of our case study indicates that webloggers who are more active are more likely to get significant assistance and support. The latter aspect can lead to a virtuous cycle in the development of reciprocity.

Signaling competences. The more an individual weblogger engages in knowledge practices, such as improving the Longhorn software, the higher will be the chance of attracting attention. For example, our data show that intensive webloggers are increasingly included in blogrolls and permalinks. In addition, several comments provide evidence that they are conscious about this effect. They seem to understand that writing intriguing entries and commentaries is an effective way to signal personal competence. Posting and commenting on practices can be interpreted as gift giving, which in anthropological research is known to establish status (Malinowski, 1922; Mauss, 1990). This contention can be illustrated by the following comment of one blogger, who reflects about his usage of weblogs as follows:

> *It is how I show my expertise in upcoming technology and distinguish my name.*

We propose that based on the signaling of competence, knowledge workers emphasize the attraction of recognition from peers. This becomes a precondition for stimulating extrinsic motivation by signaling competence. One weblogger stated in his weblog:

> *Seriously, I am pretty lucky. I have developers coming into my office asking me to dream up cool ways to show off their features in the platform that I know most designers would wet their pants for (for instance compositing multiple broadcast quality video elements with alpha masking ... mmm ... yummy).*

Research on OSS gives evidence that signaling competence and building up reputation are related to monitoring mechanisms of peer review (Raymond, 2001). Such mechanisms allow knowledge exchange through minimal transaction costs and facilitate spontaneous input of knowledge (Langlois and Robertson, 1995). In contrast to a hierarchical organization of knowledge, spillovers are not neglected (Garzarelli, 2004) by the decentralized coordination within a blogosphere. The technological features of the weblog technology support to a large extent the possibility to signal competence and build up reputation. First, the time-stamped display of the conversations in reverse chronological order provides the opportunity to trace back the origin of individual contributions. Therefore, outstanding knowledge becomes attributable to single webloggers, with the secondary effect of making the texture of organizational knowledge more visible. The latter aspect of signaling might be especially relevant in a corporate context (Lerner and Tirole, 2002; Lakhani and von Hippel, 2003). Second, weblog technology provides mechanisms to recognize individual webloggers for their perspective-making behavior. Our case study offers evidence for the common practice that outstanding knowledge sharing will lead to entries in blogrolls and permalinks. In consequence, this leads to prominent positions in the networked blogosphere. If people with a high reputation write about their knowledge practices, most members of the blogosphere will follow the conversation. This fact cannot only be traced back to technical features, such as RSS, but also to social status. Therefore, we suggest that communication of these prominent webloggers can be labeled as an "exhortation" to knowledge sharing (Gherardi, 2006: 182). Simultaneously, the powerful role of these individuals in the network finally reveals the social structure of the blogosphere.

Getting support. Looking at the MLB, we could assert that the activity of participation varies between the webloggers. If we understand the activity of participation in terms of number of written posts and written comments, we identify that the more the webloggers actively engage themselves, the more comments they are likely to get, that is, the more concrete assistance and support they can expect. Those webloggers who contributed more intensively were also contacted more frequently, an observation that is in line with previous OSS studies (e.g. Kollock, 1999; Mikkonen *et al.*, 2007). Thus, it seems as if webloggers who have a strong attachment to the blogosphere will be more likely to help others when they experience problems or have questions. In turn, they also receive considerably more feedback in comparison to less active MLB members. Consequently, the more actively a weblogger engages in weblogging, the better the chances are to learn new perspectives. To put it differently, one can assert a high degree of reciprocity within the MLB (Constant *et al.*, 1996). Moreover, the reciprocity leads to the emergence of subgroups or clusters around specific knowledge topics and practices in the blogosphere.

Taking a closer look at our data, we can describe reciprocity as being generalized (Gouldner, 1960), which seems to be altruistic at first glance. This is due to the fact that there are altruistic and prosocial webloggers who seek to augment the welfare of other people, or rather the development of the Longhorn software (McLure Wasko and Faraj, 2000, 2005), obviously without expecting any short-term or direct response in the form of feedback or assistance. Moreover, these webloggers do not post very frequently, so they cannot expect to attain reputation. This assumption is best exemplified by the comment of an MLB participant who stated that he has "important information to share with the community", or another weblogger who simply voiced: "I like to divulge". Therefore, talking about knowledge practices can be altruistic as it is not solely meant for one's own sake, but provides potentially relevant information for co-workers and/or other developers:

> *There's been a lot of buzz surrounding XAML and other XML GUI languages. Other bloggers have been blogging to death about Avalon, XAML, XUL, SVG, WVG and other technologies. Here are some links to articles and documents I compiled on this topic [...].*

However, we have to notice that there can be additional explanations for what might at first appear to be altruistic behavior. For instance, the writing on individual practices can work as a mechanism of individual reflective learning. Thus, weblogging can be ascertained as getting support for individual reflective learning:

> *"I learn and gain knowledge by talking things through, an inconvenience at work where I'm surrounded by introverts! I use my blog to work through new ideas and test new theories—coming to conclusions. I need it." or "It combines reflection, publishing and networking"*

We interpret this statement of an interviewee in the sense of a reflective learning process that helps webloggers to gain new insights about themselves and their knowledge work. Therefore, weblogging opens the possibility for professional development and helps to define problems and consider possible solutions (Schön, 1991).

Conclusion

The main topic of this paper is the impact weblog technology can exert upon an innovation process by means of knowledge sharing within a blogosphere. As for our analysis, both, contributions to theory and practice can be identified. In addition, we discuss limitations of our research that allude to suggestions for future research.

Contribution to Theory and Practice

The objective of this paper was to investigate the reasons for enabling a virtual community that interconnects employees and lead users to thrive. By elucidating the relevant motivational stimuli, we are able to identify the framing factors that foster intrinsic and extrinsic motivation. Thereby, our research makes important contributions to innovation-related research with regards to three aspects.

First, phenomena from the field of open source software appear to be valid in this for-profit realm, since not only Microsoft employees but also lead users contributed their knowledge voluntarily (in a similar vein: Harhoff *et al.*, 2003; Kristensson *et al.*, 2004; McLure Wasko and Faraj, 2005; May, personal communication, 28 June 2007). Whereas oftentimes employee–lead user exchanges in innovation processes fail from a managerial stance due to concerns regarding intellectual property or the controllability of employees, and—from a developers' stance—Microsoft is avoided due to it being a for-profit corporation, and in the case of this collaborative corporate blogospheres, the potential benefits seem to outweigh the potential risks. Therefore, our study augments previous research that argues that firms can unlock existing communities outside the boundaries of the firm (Dahlander and Wallin, 2006). We were able to show that firms can establish such "outside" communities and benefit from the knowledge and innovative ideas of users (Ofek and Sarvary, 2001; Jeppesen and Frederiksen, 2006).

Second, the reasons for the voluntary and productive participation originate in our case from the distinctive characteristics of weblogs, herein hooking up to notions from the OSS arena (e.g. Franke and von Hippel, 2003; Baldwin and Clark, 2006). Revival of rather tedious intra- as well as extraorganizational IT-based communication (e.g. conventional intranets) appears to be attainable by means of weblogs.

Third, this study also elucidates in detail as to how intrinsic as well as extrinsic motivation are crucial for engaging lead users in innovation processes. We claim that intrinsic motivation can be fostered by weblog technology, an observation we infer from the fact that participants are allowed to air their own voices freely in a trustful collaborative surrounding (in a similar vein: Jarvenpaa and Leidner, 1999; Stewart, 2003; Szulanski *et al.*, 2004). As for extrinsic motivation, participants are deemed to be motivated by the prospect of enhancing their own reputation within the community.

Limitations and Options for Future Research

A critical re-examination of our analysis makes us identify several issues for future research on innovation management within online settings. One potential shortcoming concerns the nature of the MLB, which constitutes a special blogosphere since all participants possess a high degree of IT-related knowledge. Bearing this in mind, the explanatory power might be restricted to knowledge-intensive innovation processes. Hence, in blogospheres that center on other products or services, other weblogging patterns might be observable (in a similar vein: Chesbrough and Teece, 2002).

Another aspect worth considering is related to the fact that the blogosphere at stake was run by a Microsoft employee. This person was labeled "technical evangelist" and was officially instructed to spread positive word-of-mouth for Microsoft Longhorn and was supposed to attract MVPs and to engage fellow colleagues in the joint development of the software. In addition, the MLB has to be put into perspective insofar as the blogosphere consisted of 36 active webloggers. This is only a small fraction of the overall Microsoft Vista development team, which consists of approximately 4,000 developers, albeit an important one due to the lead user characteristics of the MVPs (May, personal communication, 22 June 2007).

Moreover, our analysis could be extended by scrutinizing the kind of innovation processes that are amenable for weblogging. Herein, our propositions provide a starting point, insofar as they reveal the utility of weblogs for IT-based innovation. Nevertheless, this might not hold true for other work settings. By means of these considerations, further research endeavors can likewise be identified.

In addition to the avenues for future research mentioned above, weblogs might not only serve well as a communication device, but also as a medium for knowledge and information storage in the course of innovation processes. Herein, weblogs might also inherit a potential for reflection concerning one's own work and therefore individual learning. Bearing these thought-provoking possibilities in mind, weblogs are deemed to be worth further research and may provide people and organizations in knowledge-intensive environments practical insights that foster their own work.

References

Agarwal, R. and Karahanna, E. (2000) Time flies when you're having fun: cognitive absorption and beliefs about information technology usage, *MIS Quarterly*, 24(4), pp. 665–694.

Amabile, T. M. (1997) Motivation creativity in organizations: on doing what you love and loving what you do, *California Management Review*, 40, pp. 39–58.

Amin, A. and Cohendet, P. (2000) Organisational learning and governance through embedded practices, *Journal of Management and Governance*, 4, pp. 93–116.

Ardichvili, A., Page, V. and Wentling, T. (2003) Motivation and barriers to participation in virtual knowledge-sharing communities of practice, *Journal of Knowledge Management*, 7(1), pp. 64–77.

Bagozzi, R. and Dholakia, U. (2006) Open source software user communities: a study of participation in Linux user groups, *Management Science*, 52(7), pp. 1099–1115.

Baldwin, C. and Clark, K. (2006) The architecture of participation: does code architecture mitigate free riding in the open source development model?, *Management Science*, 52(7), pp. 1116–1127.

Bandura, A. (2000) *Self-efficacy* (New York: Freeman).

Benabou, R. and Tirole, J. (2003) Intrinsic and extrinsic motivation, *Review of Economic Studies*, 70, pp. 489–520.

Benkler, Y. (2002) Coase's penguin, or, Linux and the nature of the firm, *The Yale Law Journal*, 112, pp. 369–446.

Bergquist, M. and Ljungberg, J. (2001) The power of gifts: organizing social relationships in open source communities, *Information Systems Journal*, 11, pp. 305–320.

Bergquist, M., Ljungberg, J. and Snis, U. (2001) Practising peer review in organisations: a qualifier for knowledge dissemination and legitimisation, *Journal of Information Technology*, 16, pp. 99–112.

Brand, A. G. and Leckie, P. A. (1988) The emotions of professional writers, *The Journal of Psychology*, 122(5), pp. 421–439.

Chesbrough, H. W. and Teece, D. J. (2002) Organizing for innovation: when is virtual virtuous?, *Harvard Business Review*, 80(8), pp. 127–135.

Chesbrough, H., Vanhaverbeke, W. and West, J. (2006) *Open Innovation—Researching a New Paradigm* (Oxford: Oxford University Press).

Ciborra, C. and Patriotta, G. (1998) Groupware and teamwork in R and D: limits to learning and innovation, *R&D Management*, 28(1), pp. 43–52.

Constant, D., Kiesler, S. and Sproull, L. (1994) What's mine is ours, or is it? A study of attitudes about information sharing, *Information Systems Research*, 5(4), pp. 400–421.

Constant, D., Sproull, L. and Kiesler, S. (1996) The kindness of strangers: the usefulness of electronic weak ties for technical advice, *Organization Science*, 7(2), pp. 119–135.

Csikszentmihalyi, M. (1975) *Beyond Boredom and Anxiety* (San Francisco: Jossey-Bass).

Csikszentmihalyi, M. (1997) *Finding Flow: The Psychology of Engagement with Everyday Life* (New York: Basic Books).

Csikszentmihalyi, M. and LeFevre, J. (1989) Optimal experience in work and leisure, *Journal of Personality and Social Psychology*, 56(5), pp. 815–822.

Dafermos, G. N. (2003) *Blogging the Market. How Weblogs are Turning Corporate Machines into Real Conversations.* Available at http://opensource.mit.edu/papers/dafermos3.pdf (accessed 14 September 2004).

Dahlander, L. (2005) Appropriation and appropriability in open source software, *International Journal of Innovation Management*, 9(3), pp. 259–285.

Dahlander, L. and McKelvey, M. (2005) Who is not developing open source software? Non-users, users, and developers, *Economics of Innovation and New Technology*, 14(7), pp. 617–635.

Dahlander, L. and Wallin, M. W. (2006) A man on the inside: unlocking communities as complementary assets, *Research Policy*, 35(8), pp. 1243–1259.

Davenport, T. and Prusak, L. (1998) *Working Knowledge: How Organizations Manage What They Know* (Boston: Harvard Business School Press).

Davenport, T. H. (1997) *Information Ecology* (Oxford: Oxford University Press).

Deci, E. L. (1975) *Intrinsic Motivation* (New York: Plenum Press).

Deci, E. L. and Ryan, R. M. (1985) *Intrinsic Motivation and Self-Determination in Human Behavior* (New York: Plenum Press).

Deci, E. L., Koestner, R. and Ryan, R. M. (1999) A meta-analytic review of experiments examining the effects of extrinsic rewards on intrinsic motivation, *Psychological Bulletin*, 125, pp. 627–668.

Emerson, H. (1998) Flow and occupation: a review of the literature, *Canadian Journal of Occupational Therapy*, 65(1), pp. 37–44.

Franke, N. and von Hippel, E. (2003) Satisfying heterogeneous user needs via innovation toolkits: the case of Apache security software, *Research Policy*, 32, pp. 1199–1215.

Frey, B. (1997) *Not Just for the Money—An Economic of Personal Motivation* (Cheltenham: Edward Elgar).

Frey, B. and Oberholzer-Gee, F. (1997) The cost of price incentives: an empirical analysis of motivation crowding-out, *American Economic Review*, 87(4), pp. 746–755.

Gagliardi, P. (1996) Exploring the aesthetic side of organizational life, in: S. Clegg, C. Hardy & W. Nord (Eds), *Handbook of Organization Studies*, pp. 566–580 (London: Sage).

Garzarelli, G. (2004) Open source software and the economics of organization, in: J. Birner & P. Garrouste (Eds), *Markets, Information and Communication*, pp. 47–62 (London and New York: Routledge).

Ghani, J. A. and Deshpande, S. P. (1994) Task characteristics and the experience of optimal flow in human–computer interaction, *The Journal of Psychology*, 128(4), pp. 381–391.

Gherardi, S. (2003) Knowing as desiring: mythic knowledge and the knowledge journey in communities of practitioners, *Journal of Workplace Learning*, 15(7/8), pp. 352–358.

Gherardi, S. (2006) A journey beyond institutional knowledge: Dante's reading of the Odyssey, in: P. Gagliardi & B. Czarniawska (Eds), *Management Education and Humanities*, pp. 174–194 (Cheltenham: Edward Elgar).

Glaser, B. G. and Strauss, A. L. (1967) *The Discovery of Grounded Theory* (Chicago: Aldine).

Gouldner, A. (1960) The norm of reciprocity: a preliminary statement, *American Sociological Review*, 25(2), pp. 161–178.

Hannemyr, G. (1999) Technology and pleasure: considering hacking constructive, *First Monday*, 4(2). Available at www.firstmonday.org/issues/issue4_2/gisle/ (accessed 14 March 2007).

Harhoff, D., Henkel, J. and von Hippel, E. (2003) Profiting from voluntary information spillovers: how users benefit from freely revealing their innovations, *Research Policy*, 32(10), pp. 1753–1769.

Hars, A. and Ou, S. (2002) Working for free? Motivations for participating in open source projects, *International Journal of Electronic Commerce*, 6(3), pp. 25–39.

Herring, S., Scheidt, L., Bonus, S. and Wright, E. (2005) Weblogs as a bridging genre, *Information, Technology & People*, 18(2), pp. 142–171.

Hertel, G., Niedner, S. and Hermann, S. (2003) Motivation of software developers in open source projects: an Internet-based survey of contributors to the Linux kernel, *Research Policy*, 32(7), pp. 1159–1177.

Himanen, P. (2001) *The Hacker Ethic and the Spirit of the Information Age* (New York: Random House).

Hoffman, D. L. and Novak, T. P. (1996) Marketing in hypermedia computer-mediated environments: conceptual foundations, *Journal of Marketing*, 60(7), pp. 50–68.

Holmström, B. (1999) Managerial incentive problems: a dynamic perspective, *Review of Economic Studies*, 66(1), pp. 169–182.

Jarvenpaa, S. and Leidner, D. (1999) Communication and trust in global virtual teams, *Organization Science*, 10(6), pp. 791–815.

Jeppesen, L. and Frederiksen, L. (2006) Why do users contribute to firm-hosted user communities? The case of computer-controlled music instruments, *Organization Science*, 17(1), pp. 45–63.

Kirk, J. and Miller, M. L. (1986) *Reliability and Validity in Qualitative Research* (Newbury Park, CA: Sage).

Koch, S. and Schneider, G. (2002) Effort, cooperation and coordination in an open source software project: GNOME, *Information Systems Journal*, 12(1), pp. 27–42.

Kollock, P. (1999) The economics of online cooperation: gifts and public goods in the cyberspace, in: M. Smith & P. Kollock (Eds), *Communities in Cyberspace*, pp. 220–242 (London: Routledge).

Kristensson, P., Gustafsson, A. and Archer, T. (2004) Harnessing the creative potential among users, *Journal of Product Innovation Management*, 21, pp. 4–14.

Kuk, G. (2006) Strategic interaction and knowledge sharing in the KDE developer mailing list, *Management Science*, 52(7), pp. 1031–1042.

Lakhani, K. and Wolf, R. (2005) Why hackers do what they do: understanding motivation and effort in free/open source software projects, in: J. Feller, B. Fitzgerald, S. Hissam & K. Lakhani (Eds), *Perspectives on Free and Open Source Software*, pp. 3–22 (Cambridge, MA: MIT Press).

Lakhani, K. R. and von Hippel, E. (2003) How open source software works: "free" user to user assistance, *Research Policy*, 32(6), pp. 923–943.

Lane, C. and Probert, J. (2007) The external sourcing of technological knowledge by US pharmaceutical companies: strategic goals and inter-organizational relationships, *Industry and Innovation*, 14(1), pp. 5–25.

Langlois, R. N. and Robertson, P. L. (1995) *Firms, Markets and Economic Change: A Dynamic Theory of Business Institutions* (London: Routledge).

Lave, J. and Wenger, E. (1991) *Situated Learning. Legitimate Peripheral Participation* (Cambridge: Cambridge University Press).

Leiponen, A. (2005) Organization of knowledge and innovation: the case of Finnish business services, *Industry and Innovation*, 12(2), pp. 185–203.

Lerner, J. and Tirole, J. (2001) The open source movement: key research questions, *European Economic Review*, 45(4–6), pp. 819–826.

Lerner, J. and Tirole, J. (2002) Some simple economics of open source, *The Journal of Industrial Economics*, 50(2), pp. 197–234.

Levy, S. (1984) *Hackers. Heroes of the Computer Revolution* (London: Anchor Press).

Lin, Y. (2003) The institutionalization of hacking practices, *Ubiquity*, 4(4), pp. 18–24.

Loasby, B. J. (1998) The organisation of capabilities, *Journal of Economic Behavior & Organization*, 35(2), pp. 139–160.

Loasby, B. J. (2001) Time, knowledge and evolutionary dynamics: why connections matter, *Journal of Evolutionary Economics*, 11(4), pp. 393–412.

Malinowski, B. (1922) *Argonauts of the Western Pacific: An Account of Native Enterprise and Adventure in the Archipelagoes of Melanesian New Guinea* (London: Routledge).

Marshall, A. (1920) *Principles of Economics* (London: Macmillan).

Mauss, M. (1990) *The Gift: The Form and Reason for Exchange in Archaic Societies* (New York: Norton).

May, L. (2007) Microsoft Germany spokesperson, personal communication, 22 June.

May, L. (2007) Microsoft Germany spokesperson, personal communication, 28 June.

McArthur, R. and Bruza, P. (2001) Representation of knowledge in online communities. Paper presented at the 2nd ECSCW Workshop on Community Knowledge, Bonn.

McLure Wasko, M. and Faraj, S. (2000) "It is what one does": why people participate and help others in electronic communities of practice, *Journal of Strategic Information Systems*, 9(2/3), pp. 155–173.

McLure Wasko, M. and Faraj, S. (2005) Why should I share? Examining social capital and knowledge contribution in electronic networks of practice, *MIS Quarterly*, 29(1), pp. 35–57.

Microsoft (2005) *Most Valuable Professional—Letter from Lori, Sean and Anthony*. Available at http://mvp.support.microsoft.com/MVPINTRO (accessed 9 May 2005).

Mikkonen, T., Vadén, T. and Vainio, N. (2007) The Protestant ethic strikes back: open source developers and the ethic of capitalism, *First Monday*, 12(2). Available at http://firstmonday.org/issues/issue12_2/mikkonen/index.html (accessed 14 March 2007).

Miles, M. B. and Huberman, A. M. (1994) *Qualitative Data Analysis* (London: Sage).

Moglen, E. (1999) *Anarchism Triumphant: Free Software and the Death of Copyright.* Available at http://www.firstmonday.org/issues/issue4_8/moglen/index.html (accessed 17 May 2007).

Moody, G. (2002) *Rebel Code: Linux and the Open Source Revolution* (Philadelphia, PA: Perseus Books Group).

Moon, J. Y. and Sproull, L. (2000) Essence of distributed work: the case of the Linux kernel, *First Monday*, 5(11). Available at http://firstmonday.org/issues/issue5_11/moon/index.html (accessed 21 November 2007).

Nakamura, J. and Csikszentmihalyi, M. (2002) The concept of flow, in: S. J. Lopez & C. R. Snyder (Eds), *Handbook of Positive Psychology*, pp. 89–105 (Oxford: Oxford University Press).

Nohara, H. and Verdier, E. (2001) Sources of resilience in the computer and software industries in France, *Industry and Innovation*, 8(2), pp. 201–220.

Nonaka, I. (1994) A dynamic theory of organizational knowledge creation, *Organization Science*, 5(1), pp. 14–37.

Ofek, E. and Sarvary, M. (2001) Leveraging the cluster base: creating competitive advantage through knowledge management, *Management Science*, 47(11), pp. 1441–1456.

O'Mahony, S. (2003) Guarding the commons: how community managed software projects protect their work, *Research Policy*, 32(7), pp. 1179–1198.

Osterloh, M. and Frey, B. (2000) Motivation, knowledge transfer, and organizational forms, *Organization Science*, 11(5), pp. 538–550.

Osterloh, M., Frost, J. and Frey, B. (2002) The dynamics of motivation in new organizational forms, *International Journal of the Economics of Business*, 9(1), pp. 61–77.

Pace, S. (2004) A grounded theory of the flow experiences of Web users, *International Journal of Human–Computer Studies*, 60(3), pp. 327–363.

Raub, S. and von Wittich, D. (2004) Implementing knowledge management, *European Management Journal*, 22(6), pp. 714–728.

Raymond, E. (2001) *The Cathedral & the Bazaar. Musings on Linux and Open Source by an Accidental Revolutionary* (Sebastopol, CA: O'Reilly & Associates).

Rettie, R. (2001) An exploration of flow during Internet use, *Internet Research: Electronic Networking Applications and Policy*, 11(2), pp. 103–113.

Roberts, J., Hann, I. and Slaughter, S. (2006) Understanding the motivations, participation, and performance of open source software developers: a longitudinal study of the Apache projects, *Management Science*, 52(7), pp. 984–999.

Schön, D. A. (1991) *The Reflective Practitioner: How Professionals Think in Action* (Aldershot: Ashgate Arena).

Sheldon, K. M. and Elliot, A. J. (1998) Not all personal goals are personal: comparing autonomous and controlled reasons for goals as predictors of effort and attainment, *Personality and Social Psychology Bulletin*, 24(5), pp. 94–105.

Simon, B., Loewy, M., Stürmer, S., Weber, U., Freytag, P., Habig, C., Kampmeier, C. and Spahlinger, P. (1998) Collective identification and social movement participation, *Journal of Personality and Social Psychology*, 74, pp. 646–658.

Sorenson, O. and Singh, L. (2007) Science, social networks and spillovers, *Industry and Innovation*, 14(2), pp. 219–238.

Stallman, R. M. (2001) *Free Software: Freedom and Cooperation.* Available at http://www.gnu.org/events/rms-nyu-2001-transcript.txt (accessed 21 January 2007).

Stewart, K. (2003) Trust transfer on the World Wide Web, *Organization Studies*, 14(1), pp. 5–17.

Strati, A. and de Monthoux, P. G. (2002) Organizing aesthetics, *Human Relations*, 55(7), pp. 755–766.

Strauss, A. and Corbin, J. (1990) Grounded theory research: procedures, canons and evaluative criteria, *Qualitative Sociology*, 13(1), pp. 3–21.

Szulanski, G., Cappetta, R. and Jensen, R. (2004) When and how trustworthiness matters: knowledge transfer and the moderating effect of causal ambiguity, *Organization Science*, 15(5), pp. 600–613.

Torvalds, L. and Diamond, D. (2002) *Just for Fun: The Story of an Accidental Revolutionary* (New York: HarperCollins).

Turvani, M. (2001) Microfoundations of knowledge dynamics within the firm, *Industry and Innovation*, 8(3), pp. 309–323.

von Hippel, E. (1989) New product ideas from "lead users", *Research Technology Management*, 32, pp. 24–27.

von Krogh, G., Spaeth, S. and Lakhani, K. (2003) Community, joining, and specialization in open source software innovation: a case study, *Research Policy*, 32(7), pp. 1217–1241.

Walther, J. B. and Tidwell, L. C. (1995) Nonverbal cues in computer-mediated communication, and the effect of chromenics on relational communication, *Journal of Organizational Computing & Electronic Commerce*, 5(4), pp. 355–378.

Weber, S. (2004) *The Success of Open Source* (Cambridge, MA: Harvard University Press).

Yin, R. (2003) *Case Study Research* (Thousand Oaks, CA: Sage).

Zeitlyn, D. (2003) Gift economies in the development of open source software: anthropological reflections, *Research Policy*, 32(7), pp. 1287–1291.

Zwick, R. (1988) Another look at interrater agreement, *Psychological Bulletin*, 103, pp. 374–378.

Getting Clear About Communities in Open Innovation

JOEL WEST & KARIM R. LAKHANI

ABSTRACT Research on open source software, user innovation and open innovation have increasingly emphasized the role of communities in creating, shaping and disseminating innovations. However, the comparability of such studies has been hampered by the lack of a precise definition of the community construct. In this paper we review prior definitions (implicit and explicit) of the community construct, and other suggestions for future research.

An important development in the past decade of innovation studies has been the recognition of the role of communities outside of the boundaries of firms in creating, shaping and disseminating technological and social innovations. While the dominant role of users in creating functionally novel innovations was established 20 years ago by von Hippel (1988), the advent of open source software communities has highlighted the important role of communities in the innovation process. The relative success of the open source movement, and its novel collaboration, problem solving and intellectual property practices, have also focused the attention of innovation scholars on the "community" phenomenon and its implications for innovation theory and practice.

Concurrent with the rise in studies of community in the innovation process, there has been a parallel and distinct body of literature focused on "open innovation" (Chesbrough, 2003). This literature has tended to focus on interfirm cooperation and the development of an ecosystem of firms, sharing technologies and trading intellectual property, within a given industry or sector (cf. West *et al.*, 2006). However, non-firm actors such as communities are rarely to be found in the recent writings on open innovation. However, community-based innovation by its nature takes place outside the boundaries of the firm, which fits Chesbrough's definition (2003) of open innovation. Thus communities and their role in the innovation process both fit within and offer an opportunity to extend the firm-centric concept of open innovation developed by Chesbrough and his colleagues (Chesbrough, 2003; Chesbrough *et al.*, 2006; Gassmann, 2006).[1]

The papers in this Special Issue are an attempt to explicitly link the role of communities in open innovation. But what is a community? If we are considering a community as a level of analysis then we need a precise definition of "community". In scholarly research, the definition of community may be explicit, implicit or phenomenological, as illustrated by the papers of this Special Issue. The definition may be explicitly stated in terms of prior research, as West and O'Mahony (this issue) do in citing technical communities, Kaiser and Müller-Seitz (this issue) do with communities of practice or Finotto and Di Maria (this issue) do for user communities. In other cases, communities are implicitly defined by citation of prior research, as when Langlois and Garzarelli (this issue) cite Allen (1983). Still other papers use the term to refer to the phenomenon they are studying—as do den Besten and Dalle (this issue)—which helps others studying the same phenomenon but unduly limits the applicability of their work to other phenomena.

These papers and others suggest a welter of overlapping literatures and terms: innovation communities, knowledge producing communities, online communities, scientific communities, technical communities, user communities, virtual communities or communities of practice. That doesn't even include the disparate uses of "community" in sociology, where Brint (2001) recounts that some 100 different definitions have been used.

Here we do not attempt to impose a single definition of "community" applicable to all innovation studies. Instead, we encourage researchers to explicitly articulate the theoretical and phenomenological boundaries of their use of the term, both to build upon earlier scholarship and to enable subsequent scholars to build on them.

To advance such precision and enable researchers to better fit their community innovation research to the broader community of innovation researchers, we contrast some of the major similarities and differences across the prior definitions, noting ambiguities and unresolved issues in such definitions. We also consider questions of how such communities are constituted and operate. From that, we suggest opportunities for future research on community innovation.

Defining the Community Construct in Open Innovation

While there are many types of communities, we are only interested in those involved in creating innovation outside the boundaries of the firm. We thus bound our scope in two ways. First, by building upon the definition of Gläser (2001), we consider a community to be a voluntary association of actors, typically lacking in a priori common organizational affiliation (i.e. not working for the same firm) but united by a shared instrumental goal—in this case, creating, adapting, adopting or disseminating innovations.

[1] In our respective work on open source software over the past 8 years, we have encountered many cases where both scholars and industry participants questioned whether all open source software development qualifies as "innovation" rather than low-cost reimplementation of existing technologies. There are enough examples of new technologies being pioneered in open source (such as the Apache webserver)—plus all the well-documented user innovation work of Eric von Hippel and his colleagues—that here we focus on communities without regard to which ones are truly "innovative".

Secondly, in this review we focus on innovations that are explicitly brought to market or widespread use by commercial actors. While some community innovations have a purely non-profit motivation and thus do not overlap the definition of open innovation (West and Gallagher, 2006), many of the principles of community innovation described below would apply to non-profit actors as well.

Communities of What?

A fundamental question about communities is what is the level of analysis for the community's constituent members. In sociology communities are assumed to comprise a group of individuals. In innovation studies, we certainly have examples of communities containing only self-governing groups of individuals, as with open source software (Lerner and Tirole, 2002; O'Mahony, 2003).

More often, for economically valuable innovations, firms are involved either directly or indirectly. As von Hippel (2005: 96) explicitly says, "Innovation communities can have users and/or manufacturers as members and contributors." Firm participation can occur through direct sponsorship of staff and their efforts in communities. For example, Lakhani and Wolf (2005) have shown that 40 percent of contributors to open source software projects are "paid" to participate. Outside of open source, in communities developing important technical standards, the direct interactions involve individuals, but the individual actors are formally or informally representing their corporate parents, as can be seen from the subsequent corporate actions (e.g. Rosenkopf *et al.*, 2001; Fleming and Waguespack, 2007).

However, does the "community" concept also encompass groups of (only) firms interacting with each other? Lynn *et al.* (1996) refer to an "innovation community", while von Burg and Kenney (2003: 354) define community as "a cluster of autonomously acting firms agreeing to be bound by the constraints of a standard". Such cooperating firms might be aligned through formal institutions—as with the Eclipse community that West and Gallagher (2006) liken to an industrial R&D consortium—or they might be coordinated through market mechanisms across a value network (Vanhaverbeke, 2006). But are these, in fact, "communities", or instead better described as a value chain, value network, ecosystem or industry segment?

Powell (1990) was among the first to notice that a network of organizations is explicitly a distinct interorganizational form from markets and hierarchies. Networks play a central role in organizing the open innovation activities of multiple cooperating firms (see Vanhaverbeke, 2006 for a review). Certainly there are overlaps in the definition between networks and communities, such as Powell's emphasis (1990) on reputation, norms of reciprocity and mutually beneficial interactions. Given these overlaps, are communities of firms synonymous with interorganizational networks? Are they a proper superset (or subset)? Do the two constructs have differing domains or theoretical predictions?

Firms and Communities

An area of increasing interest is how firms choose to interact with communities, with research that links a firm's strategy to the organization and motivation of the community. Such firms may occupy a privileged position in the community—usually communities of their own creation—or may instead interact with existing communities.

In some cases, firms are merely one of the constituent members in a community that contains multiple types of actors. A well-known example of such a heterogeneous mix of actors is the Linux kernel community, with individuals, firms and two (recently combined to one) non-profit foundations.

More commonly, firms play a central role in creating and organizing innovation communities. Often, a single firm holds a privileged role in which it seeks to guide or control the community towards achieving firm-level objectives. Such control of open source communities has attracted increasing managerial and theoretical interest: Välimäki (2003) was the first to look at the impact of firm control (of IP terms) upon an external community, while West and O'Mahony (this issue) establish more broadly the mechanisms by which firms both control and seek participation from open source communities. Such sponsored open source communities are becoming more common as the successful model of MySQL (and its acquisition for $1 billion in early 2008) becomes widely known and emulated.

Such firm-sponsored communities are not limited in open source software. Firms sponsor communities to develop modifications to other information goods, such as music synthesizers or video games (Jeppesen and Frederiksen, 2006; West and Gallagher, 2006). Finotto and Di Maria (this issue) show how these principles can be applied to fashion and motorcycle equipment.

However, a focus on successful firm interactions in these studies would understate the likelihood of firms and communities being able to find a common interest. Firms may be held at arm's length from communities if a self-governing group of individuals wish to insulate themselves from corporate influence (O'Mahony, 2003). Firms that attempt to create an innovation community may fail due to lack of adopter interest, or may neglect communities if the innovation returns prove disappointing (Dahlander and Magnusson, 2005). Conversely, firms may ignore the needs of the community, creating a demand for users to create their own companies to commercialize technology. Such user innovator startups can be found in water sports (Hienerth *et al.*, 2006), university-developed open source software (West and O'Mahony, this issue) and juvenile products (Shah and Tripsas, 2007).

Research on firms sponsoring communities may also overstate the intentionality of firm interactions with communities. What is the relative frequency (or economic importance) of such intentional action, versus a more passive, peripheral or responsive approach? As von Hippel (1988) reminds us, firm use of community innovations can also be serendipitous—an emergent rather than planned strategy.

Intra-community Interactions

Fundamentally it is important to be clear about what interactions constitute the basis for a group of individuals and/or firms to be in a community. Is a community more than just a grouping of individuals (or other actors)? Is there a community if there's no sense of community (cf. Corlett, 1989)? In particular, what role do interactions between members play in defining communities in the open innovation process?

The communities of practice literature stipulates that knowledge transfer between community members is the key prerequisite to organizational learning, particularly for communities subsumed within an organization. The only reason we remark on the community of practice exemplified by Brown and Duguid's copier technicians (1991) is that their interactions allow them to share, refine, pool and disseminate best practice for their work responsibilities. Similarly, Knorr Cetina (1999) has shown that collective problem solving and joint artifact creation form the basis of connection and community amongst the individuals and institutions engaged in high energy physics research.

Among studies related to open innovation, the intra-community linkages in some studies are more explicit than others. User innovation researchers have studied the peer-to-peer assistance in open source software (Lakhani and von Hippel, 2003) and sporting goods (Franke and Shah, 2003). For these cases, the community support facilitates adoption and use of the innovation, although only in the latter case does that adoption create revenues for a firm (such as buying a sailplane or a snowboard). Meanwhile, identification and interaction within a user community means that innovations fuel imitation and extension by other user innovators (von Hippel, 2001).

Intra-community interactions also offer the prospect of defining the boundaries of a given innovation community. The boundaries may not be sharply defined, as the differences between core and periphery may be more dramatic than the differences between peripheral members and nonmembers. However appealing such core–periphery distinctions are, they may be highly dependent on both definitional and methodological choices made by researchers as to what constitutes community membership. From a sample of 116 SourceForge projects, Crowston *et al.* (2006) found different definitions of core membership by using formal responsibilities, code contributions and network interactions.

This still begs the question of user innovation *by* the community that is not *within* the community. That is to say, what if there is no intra-community interaction, for instance, users disclose their innovations to firms but not to each other—as with the scientific instruments innovations of von Hippel (1988)? Does such free revealing by a population of users—without interactions between those users—constitute community innovation? Less dramatically, if a community is held together by the firm as the only bridge between disparate members, then is the community an innovation community or merely a customer base?

Even when direct interactions exist, they are often just part of the picture. For example, research on open source communities has focused on the direct interactions between community members, such as through email interactions. However, the primary outcome (or even antecedent) of successful open source communities is not such interaction, but the joint or cumulative production of a shared information good, that is, the open source software. At the same time, other parts of the same open source community (such as user-to-user assistance) may be organized around such direct interactions, suggesting an opportunity for network analysis of the innovation flows between these two activities.

Future Research

The disparate definitions and measures provide important opportunities for studying the role of communities in innovation. Here we offer some suggestions as to some of these opportunities.

Defining the Community Construct

The study of community innovation would first require defining what a community is—either through a more extensive review of prior literature or through empirical study. While we have attempted to survey the use of communities in innovation research, we've made only passing references to the considerable sociological research on communities of individuals. Meanwhile, analyzing communities with individual and organizational members requires not only spanning levels of analysis, but also (as Stein, 1997 shows) may require spanning theories with conflicting fundamental assumptions.

Other parts of the definition—such as the boundaries of innovation communities—may only be resolved through empirical study. The largest number of empirical studies of community innovation have focused on open source communities, and such communities have influenced both the theory and practice of innovation in other communities. However, even a superficial examination suggests important differences between software communities and those that (like Wikipedia) produce other types of information goods and even physical goods. These communities may have different perspectives on membership, knowledge sharing, intellectual property and commercialization opportunities. These thus may be different in kind rather then degree, implying generalization across the phenomenon may be difficult.

As noted earlier, any definition of "community" needs to consider whether the construct applies to associations of firms (as opposed to individuals or mixed groupings). If so, how do "communities" differ from other constructs—particularly Powell's definition of a network (1990), with its open-ended interdependence and relational coordination? Are there communities within networks or networks within communities? Does any set of firms with complementary offerings constitute a community, or only those with a specific shared objective (as in the definition of von Burg and Kenney, 2003)?

Interactions With and Within Communities

Communities differ in how they interact, both in terms of differences in types of members and differences in how communities are organized. Several of the papers in this issue have considered communities of individuals that interact with a sponsoring firm, but does this generalize to more complex community forms? The interplay of charitable trusts, non-government organizations, governments, universities and corporations to develop vaccines for sub-Saharan Africa offer but one example.

If we are considering differences in community interactions, are such differences economically or managerially significant to the community innovation process? If so, we would expect to see differences in important outcomes such as innovativeness (whether quantity or originality), speed, community mortality rate or the ability of firms to commercialize innovations. Such difference might reflect direct effects or interactions between the nature and organization of the community and imply the development of more contingent theories.

Thus far, the interactions of firms and communities has tended to assume an intentional one-way flow from firms to community structure and organization. For example, West and O'Mahony (this issue) bifurcate between sponsored and autonomous communities, but to what degree do sponsored communities show self-governing traits (consistent with O'Mahony, 2007). In many sponsored communities (such as MySQL) firms tightly control code including the technical modularity that enables decentralized innovation, but what is the role of communities in enacting, negotiating or interpreting such modularity?

Finally, we've assumed that firms play a positive role in nurturing and developing communities. But what about parasitic actions by firms—that attempt to harvest the benefits of community innovation without contributing any resources to it? Conversely, communities can turn against firms and create significant economic damage and disruption (e.g. Mollick, 2005). Conflict amongst firms and communities is relatively understudied and its study may provide important insights into the community-based open innovation process.

Communities and Open Innovation

Prior research has identified the importance of individuals, firms, networks, industries and nations upon open innovation (West *et al.*, 2006). Although much of the emphasis in open innovation has been on dyadic interactions between firms, clearly communities are also an important source of innovations, innovations that have been utilized by firms as inputs to their own innovation strategies (e.g. von Hippel, 2005).

How should communities be considered in the context of these other levels of analysis? Would an examination of the role of communities in open innovation produce any different predictions than examining the role of individual members of a community? Are there important industry or national differences in the suitability of communities as a source of innovations?

Although the user innovation literature has pioneered the understanding of community innovation, as the name suggests, the emphasis has been on innovations from downstream (the customer) side of the value chain.[2] Conversely, both research and the practice of open innovation have emphasized the supply of inputs to a firm's innovation funnel. Does this gap reflect a fundamental difference in practice or just a difference in conceptualization? Is it due to differences in the assumed motivation—personal utility (scratching an itch) versus financial gain?

Such questions suggest multiple opportunities to integrate the study of innovation communities with the practice of open innovation. There is a strong managerial trend towards embracing innovation from outside the boundaries of the firm (Raynor and Panetta, 2005) and the emergence of hybrid innovation models that integrate firm and communities (Lakhani and Panetta, 2007). The challenge for scholars will be to build theories and derive empirical evidence to improve both the theory and practice of innovation.

Acknowledgement

The authors thank Wim Vanhaverbeke and the editors for their helpful suggestions.

References

Allen, R. C. (1983) Collective invention, *Journal of Economic Behavior and Organization*, 4(1), pp. 1–24.

Brint, S. (2001) Gemeinschaft revisited: a critique and reconstruction of the community concept, *Sociological Inquiry*, 19(1), pp. 1–23.

[2] As its title suggests, von Hippel (1988) takes the broadest possible view of innovation sources, explicitly including innovation by suppliers. Nearly all of the research that builds on von Hippel has focused strictly on users.

Brown, J. S. and Duguid, P. (1991) Organizational learning and communities-of-practice: toward a unified view of working, learning, and innovation, *Organization Science*, 2(1), pp. 40–57.

Chesbrough, H. (2003) *Open Innovation: The New Imperative for Creating and Profiting from Technology* (Boston: Harvard Business School Press).

Chesbrough, H., Vanhaverbeke, W. and West, J. (Eds) (2006) *Open Innovation: Researching a New Paradigm*, pp. 82–106 (Oxford: Oxford University Press).

Corlett, W. S. (1989) *Community Without Unity: A Politics of Derridian Extravagance* (Durham, NC: Duke University Press).

Crowston, K., Wei, K., Li, Q. and Howison, J. (2006) Core and periphery in free/libre and open source software team communications, *Proceedings of the 39th Annual Hawaii International Conference on System Sciences*, p. 118A.

Dahlander, L. and Magnusson, M. G. (2005) Relationships between open source software companies and communities: observations from Nordic firms, *Research Policy*, 34(4), pp. 481–493.

den Besten, M. and Dalle, J-M. (this issue) Keep it simple: a companion for Simple Wikipedia?.

Finotto, V. and Di Maria, E. (this issue) Communities of consumption and Made in Italy.

Fleming, L. and Waguespack, D. M. (2007) Brokerage, boundary spanning, and leadership in open innovation communities, *Organization Science*, 18(2), pp. 165–180.

Franke, N. and Shah, S. (2003) How communities support innovative activities: an exploration of assistance and sharing among end-users, *Research Policy*, 32(1), pp. 157–178.

Gassmann, O. (2006) Opening up the innovation process: towards an agenda, *R&D Management*, 36(3), pp. 223–226.

Gläser, J. (2001) "Producing communities" as a theoretical challenge, *Proceedings of The Australian Sociological Association* (TASA 2001), pp. 1–11.

Hienerth, C. (2006) The commercialization of user innovations: the development of the rodeo kayak industry, *R&D Management*, 36(3), pp. 273–294.

Jeppesen, L. B. and Frederiksen, L. (2006) Why do users contribute to firm-hosted user communities? The case of computer controlled music instruments, *Organization Science*, 17(1), pp. 45–63.

Kaiser, S. and Müller-Seitz, G. (this issue) Leveraging lead user knowledge in software development—the case of weblog technology.

Knorr Cetina, K. (1999) *Epistemic Cultures: How the Sciences Make Knowledge* (Cambridge, MA: Harvard University Press).

Lakhani, K. R. and Panetta, J. A. (2007) The principles of distributed innovation, *Innovations: Technology, Governance, Globalization*, 2(3), pp. 97–112.

Lakhani, K. R. and von Hippel, E. (2003) How open source software works: free user to user assistance, *Research Policy*, 32(6), pp. 923–943.

Lakhani, K. R. and Wolf, R. G. (2005) Why hackers do what they do: understanding motivation and effort in free/open source software projects, in: J. Feller, B. Fitzgerald, S. A. Hissam & K. R. Lakhani (Eds), *Perspectives on Free and Open Source Software* (Cambridge, MA: MIT Press).

Langlois, R. N. and Garzarelli, G. (this issue) Of hackers and hairdressers: modularity and the organizational economics of open-source collaboration.

Lerner, J. and Tirole, J. (2002) Some simple economics of open source, *Journal of Industrial Economics*, 52(2), pp. 197–234.

Lynn, L. H., Reddy, N. M. and Aram, J. D. (1996) Linking technology and institutions: the innovation community framework, *Research Policy*, 25(1), pp. 91–106.

Mollick, E. (2005) Tapping into the underground, *Sloan Management Review*, 46(4), pp. 21–24.

O'Mahony, S. (2003) Guarding the commons: how community managed software projects protect their work, *Research Policy*, 32(7), pp. 1179–1198.

O'Mahony, S. (2007) The governance of open source initiatives: what does it mean to be community managed?, *Journal of Management and Governance*, 11(2), pp. 139–150.

Powell, W. W. (1990) Neither market nor hierarchy: network forms of organization, in: B. M. Staw & L. L. Cummings (Eds), *Research in Organizational Behavior*, Vol. 12 (Greenwich, CT: JAI Press).

Raynor, M. E. and Panetta, J. A. (2005) A better way to R&D?, *Strategy & Innovation: A Newsletter from Harvard Business School Publishing and Innosight*, 3(2), pp. 14–16.

Rosenkopf, L., Metiu, A. and George, V. P. (2001) From the bottom up? Technical committee activity and alliance formation, *Administrative Science Quarterly*, 46(4), pp. 748–772.

Shah, S. and Tripsas, M. (2007) The accidental entrepreneur: the emergent and collective process of user entrepreneurship, *Strategic Entrepreneurship Journal*, 1(1), pp. 123–140.

Stein, J. (1997) How institutions learn: a socio-cognitive perspective, *Journal of Economic Issues*, 31(3), pp. 729–740.

Välimäki, M. (2003) Dual licensing in open source software industry, *Systemes d'Information et Management*, 8(1), pp. 63–75.

Vanhaverbeke, W. (2006) The inter-organizational context of open innovation, in: H. Chesbrough, W. Vanhaverbeke & J. West (Eds), *Open Innovation: Researching a New Paradigm*, pp. 205–219 (Oxford: Oxford University Press).

von Burg, U. and Kenney, M. (2003) Sponsors, communities, and standards: ethernet vs. token ring in the local area networking business, *Industry and Innovation*, 10(1), pp. 351–375.

von Hippel, E. (1988) *The Sources of Innovation* (New York: Oxford University Press).

von Hippel, E. (2001) Open source shows the way: innovation by and for users—no manufacturer required, *Sloan Management Review*, 42(4), pp. 82–86.

von Hippel, E. (2005) *Democratizing Innovation* (Cambridge, MA: MIT Press).

West, J. and Gallagher, S. (2006) Patterns of open innovation in open source software, in: H. Chesbrough, W. Vanhaverbeke & J. West (Eds), *Open Innovation: Researching a New Paradigm*, pp. 82–106 (Oxford: Oxford University Press).

West, J. and O'Mahony, S. (this issue) The role of participation architecture in growing sponsored open source communities.

West, J., Vanhaverbeke, W. and Chesbrough, H. (2006) Open innovation: a research agenda, in: H. Chesbrough, W. Vanhaverbeke & J. West (Eds), *Open Innovation: Researching a New Paradigm*, pp. 285–307 (Oxford: Oxford University Press).

Index

Page numbers in *Italics* represent tables.
Page numbers in **Bold** represent figures.